György Rózsa

INFORMATION: FROM CLAIMS TO NEEDS

György Rózsa

INFORMATION: FROM CLAIMS TO NEEDS

National aptitudes for international co-operation in scientific information economy

KULTURA LIBRARY OF HAS

BUDAPEST · 1988

Joint edition published by
the Library of the Hungarian Academy of Sciences
and
the Kultura Hungarian Foreign Trading Company

Editor:
dr Th. Kabdebo, Ph.D., F.L.A.

ISBN 963 7302 40 9

FOREWORD

The studies in this volume were written by a Hungarian librarian who differs from the rest at least in two respects. He has vital international experience which equips him to deal with the common and the variable features of his topic. Moreover, he is a librarian who is deeply concerned with the flow and the use of information, one of the main problems of our time. Being a researcher our librarian has a firm footing in each field: research and librarianship.

It is common knowledge that the quantity of information has increased but there is dissatisfaction in determining the use and the modes of supply. Quantitative growth enables us to extend the circle of information but it contains a warning about the necessity of selection. Registering is already difficult but it is even harder to determine the criteria of selection not to mention scientific value and social usefulness.

Our judgement should be based on a knowledge of the present position, an understanding of current economic tendencies, social movement and scientific formations. In other words our choice should look at both the present and the future. For all those reasons the supplier of information is not just a documentalist, but also a researcher.

One of the author's main point is that there is a need for an information policy capable of governing and organising the economy of information. This policy should be based on the existing institutional structure and the technical expertise together with their future possibilities. Hypotheses and research results are generally born in situations where there is a certain autonomy of research and where scientific trends and economic needs clearly express themselves. An information policy should be sensitive to trends, in other words to needs that emerge from the practical fusion of science/knowledge and practice.

The trends of international developments are differently gauged by the various research bases and units of various countries but ongoing experience enable one to make comparisons among multiple development trends and this facilitates selection. In order to formulate a national science policy it is indispensable that the policy maker should be in the possession of general and comparative information. The same is true for the practitioners of individual disciplines. The author rightly states that information centres and large libraries fulfill a research function too since the information provided enables the practitioner of a branch of science to measure its own level of development and these, in toto, enable the policy makers to be informed of the scientific capacity and its usability as it relates to the whole country. Furthermore all this gives an indication to mark further research trends and priorities.

Information-economy is particularly important for a small nation which cannot create optimal circumstances for scientific development in every branch of science so it must be informed what are the trends in more developed countries.

I think that under the circumstances we no longer speak of some sort of an intellectual claim that comprises the choice, the flow, the betterment of information but an important economic and social necessity which makes it possible for us to use the results of this branch of science in creating modernisation, transformation and reform.

The author emphasizes the economic significance of an information policy which should gain general approval since information is an important factor of economic efficiency.

Yet to form the whole picture we need to look beyond the economy towards man himself who stands in the centre of the production process as "humanisation" is an important task of the future. Despite of different social systems mankind has such global problems which are tied to keeping the peace, the environment, fighting against backwardness and barbarism in human and worldly attitudes and preserving the best in culture.

When talking about evolution we ought to think about the needs in relation to information-economics. This is the author's own view, particularly in relation to the social sciences, which is laudable more than most, as it contains his faith in a widely based culture.

Information-economics needs a national framework; here international cooperation is vital and the author (who was the director of the UN Library in Geneva and is an organiser and participant of a European system of social science information) is a promoter of these issues. His studies and essays convince us that the modes of cooperation need to be worked out together with their technical possibilities.

He examines both the advanced and the developing countries from the point of view of their information systems. This provides a contrast which shows not just the advantages and disadvantages of international cooperation but that of the information--flow and of the uncertainty of choice. He studied the problem on missions sent to developing countries and with his rational approach he endeavours to demystify the role of information systems in general and in these countries in particular.

Researcher and librarian, György Rózsa is an expert of information studies, someone who understands the possibilities and limits of information-economy because he knows that information is but the means for the creative spirit of man that moves the world forward.

With the reaffirmation of these thoughts I recommend this volume of collected studies by the Director General of the HAS to the reader. I, as minister of culture, am responsible for the cause of libraries, but beyond that, I, as researcher, am also involved with the question of information economy and information policy. My recommendation is reaffirmed by the experience of having worked together with the author at the Hungarian Academy of Sciences.

Budapest, 22 July 1987

Béla Köpeczi
Member of the Hungarian
Academy of Sciences

THE AUTHOR'S INTRODUCTION

"Knowledge is of two kinds. We know a subject ourselves, or know where we can find information upon it." (Boswell: Life of Johnson. 1775.)

"To connect the country's mind with the nation's heart." (Victor Hugo, at the French National Assembly. 1850.)

Recorded in a UNESCO document and accepted internationally is the statement: the strength of a nation lies in its raw materials, its sources of energies and information. The individual papers of this volume deal with information as a collective concept and treat the aspect of the theory and principles of *subject literature* as one of the salient topics of science and learning. From the general problems the author – given his special interests – has grouped his questions under three headings:

– The professional and social aspects of the information revolution.

– The role of international relations in the field of information and its relationship with national characteristics.

– The information infrastructure of developing countries.

As individual contributions these papers, in the main, have already been published in various collections as conference papers or as contributions to scientific and professional journals throughout the world, either in English, or in French, Russian or Hungarian. But for the first time, editorial work done for this volume gave them unity of structure.

In so far as information is a basic resource of nations, this resource is an economic category too. The economics of information requires an information economy whose social-economic mechanism is a given information policy. Consequently information, and subject information in particular, is a complex research topic which is inextricably bound together with the organisation of research, with research on learning, with informatics and with librarianship. In other words it belongs to the field of interdisciplinary research with a strong orientation towards economics.

All this makes the thought less heretical whereby a scientific organisation, like a large library or information centre should be managed similarly to a profit-making organisation. While the latter will show up a profit the former will surely yield an abstract profit in terms of intellectual efficiency.

Writings that claim to have a perspective always move along borderlines between reality and wishes – which is understandable; one cannot forecast the past nor the future. We have to be content with approximations. The scenario of the future is borne out of the present and of the analysis of the past. Consequently it has to reckon with factors that had not existed even in terms of claims at the time of writing the scenario. It is desirable that forecasts should have a tendency to stimulate claims whose basis is made up of development trends in information, the economic and cultural background of nations and of the research potentials of scientists and scholars. The realistic scenario for the future is the one that can account for *the process as claims are being transferred into needs*. Mixing the two might result in confusion, push the principle of gradual growth into the background and thwart the changes of attitudes necessary in introducing new elements in information. Yet the change is a must, together with the building of the infrastructure and the continuous induction aimed at "repackaging" the information of both macro and micro levels. Furthermore, it has to be recognised, that information will provide building blocks in future decision making. Claims turn into needs by way of a process which includes the tasks of organising scientific research – in a certain sense – in great scientific libraries. It is done by reading "correctly" those trends that are threading through scientific literature. Apart from the daily information service this "correct reading" gives specificity and extra value to the tasks of those who work with information.

Mostly in the form of essays the papers in this volume endeavour to express this view by concentrating on the three groups of questions set above. The author hopes that these writings will, in some measure, contribute to the effort whose goal has been sketched by Victor Hugo: "To connect the country's mind with the nation's heart." At the time of great changes, such as modernisation (the informatisation of society) formulating hypotheses and debatable points could, in themselves, serve the very interest of economics, of science and of technical development

* * *

The author takes pleasure in acknowledging the valuable contribution of dr. Thomas Kabdebo, Ph.D., F.L.A., the Librarian of the St. Patrick's College, Maynooth (Ireland) in the compilation of this volume.

I. THE SOCIO-PROFESSIONAL ASPECTS OF THE DEVELOPMENT OF THE SCIENTIFIC INFORMATION WITH SPECIAL REGARD TO SOCIAL SCIENCES

INFORMATION NEEDS OF THE SOCIAL SCIENCES

Synopsis

I. Distinctive features in the use of information by the two great fields of knowledge (science versus social science and humanities);

II. The "tertiary" information function or a new communication link;

III. The equal partnership (between research and information) and the discrepancy between the demand for information specialists and their shortage owing to the greater attraction of research careers;

A. The scheme of the flow of information;

B. The scheme of the flow of information with an "information officer".

Introduction

Society needs all kinds of information, irrespective of *provenance* and *form*, which may be effectively utilized within organized social activities, economic, scientific, technological and so on. What I wish to emphasize here is the *content*, the *applicability* and the *value* of information and not its channels and techniques: the latter are also significant since they make information accessible but their treatment belongs to the "technology of information".

Scientific information is a subordinate concept of the *intellectual communication system of society*. The special significance of information on special literature within the intellectual communication system of society largely depends on time, subject field and on the purpose of applications.

I. Distinctive features in the use of information by the two great fields of knowledge (science versus social science and humanities)

When investigating the practice in the use of information, one can discern distinctive features in the use of research methods and results and in the specialized literature between the two great fields of knowledge, natural sciences and engineering on the one hand, and social sciences and humanities on the other.

The *natural sciences* and engineering have, for the most part, an *experimental* character within which basic, applied and development research can be distinguished. The same may be said in the case of the *direct applicability* of their records which, in the last analysis, become essentially a *force of production*. The information and data are of an *objective* character, the *time* factor (speed) plays an important role in the acquisition of scientific information. The applicability of research results and its information thereon *do not belong* to the social system. The results of basic research are not affected by the time, but on other levels of research the durability of information rapidly *decreases*. In development research the results can be *directly measured* economically. The processing of *vast amounts* of information and data require the use of *automated methods* and techniques. The "*technical depreciation*" of a large part of information is *extremely rapid*. On account of the time factor, periodic publications, pre-prints and research reports are of decisive importance.

The social sciences have for the most part a *verbal* character, but the value of the results in the different branches again varies in practical applicability: historical scholarship and humanities, on the one hand, and concrete organizational and administrative branches on the other. The rate of the "*technical depreciation*" is slower. Information has to some extent an *ideological* aspect. The applicability of research results and its information *are related* to the social system. In general, the demand for retrospective research is greater, and beside the automated methods and techniques, *traditional* information forms, like catalogue cards, bibliographies are very significant. The information value of *books* decreases less rapidly, and among the periodical publications the weekly and daily *press* also represent a considerable reference source.

The recent tendency (in the last two decades) to satisfy users' needs for information is formulated by establishing *computer-aided systems*. It is concerned with the part of the entire body of knowledge which is subject to "technical depreciation" i.e. to *redundancy* and characterized by a vast amount of data. Considering that the bulk of data is in part readily applicable in technical and economic development, and more or less ephemeral in character, the *speed* of processing and *transfer* is the most important factor. Attempts at solving the *automation of information* follow from these factors.

In the social sciences the role of these factors is *less important*. In some fields the volume of data needs *the same treatment* of information as in natural sciences and engineering, i.e. statistics, demography, planning, but the speed-factor could be *disregarded*.

Although *no strict demarcation line* can be drawn between these two large fields of science in their use of information, the distinctive features enumerated above afford a certain basis for differentiation.

II. The "tertiary" information function or a new communication link

Concerning the information needs of a part of the social sciences requiring a great volume of data storage and retrieval, e.g. applied economics or management, the problem could be approached by "huge machine memories" as in science and technology. But *social sciences are more literature-oriented;* therefore, the approach of *textual* information needs to open new ways, new solutions *parallel* with the computer-aided programmes.

It is essential that the frequently mentioned "information explosion" should be re-formulated. The *exponential increase* of special literature has brought in its wake the *multiplication* of secondary information. An *excess* of information tends in the long run towards *too little or imprecise* information, clouding the needs of the user. In this respect, automation can help only *partially*. There is a congestion also in the flow of secondary information.

It appears, therefore, that whether the traditional or the mechanized information processing is used, it is necessary to introduce between the researcher and the information processes a *new communication link* which could be referred to, as is now the practice in specialized literature, as "information officer".

There already exist such research teams in which the exclusive task of one or two members is to provide the team with the required information on the basis of the primary and secondary sources of literature. These members of the team have the *same status* as the research members.

We might introduce a "tertiary" information function with the role of "information officer" who would analyze, on a level with researchers, documentary publication (abstracts, bibliographies, etc.) and notably bring material to the attention of *individual* researchers, *which* is relevant to their research. Another effect of the processing of the secondary information would be to draw attention to new topics and in this way *stimulate* research. The present *maximum* of the processing work of documentation would

THE SCHEME OF THE FLOW OF INFORMATION

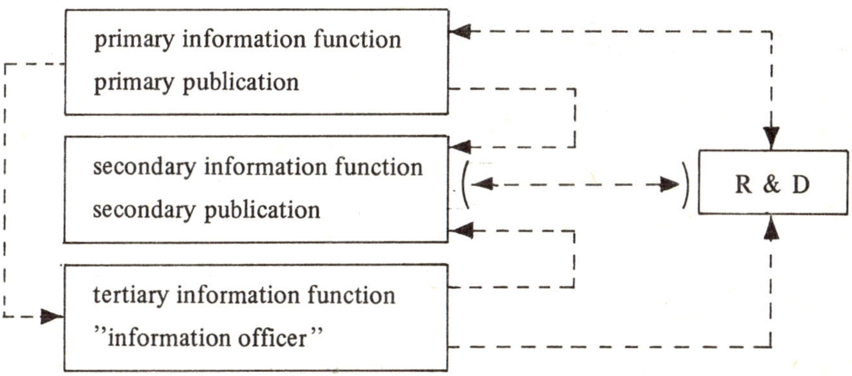

users of
- primary publication:
  R + D + documentation + "information officer"
- secondary publication:
  documentation + "information officer"
- "information officer":
  R + D

THE SCHEME OF THE FLOW OF INFORMATION WITH "INFORMATION OFFICER"

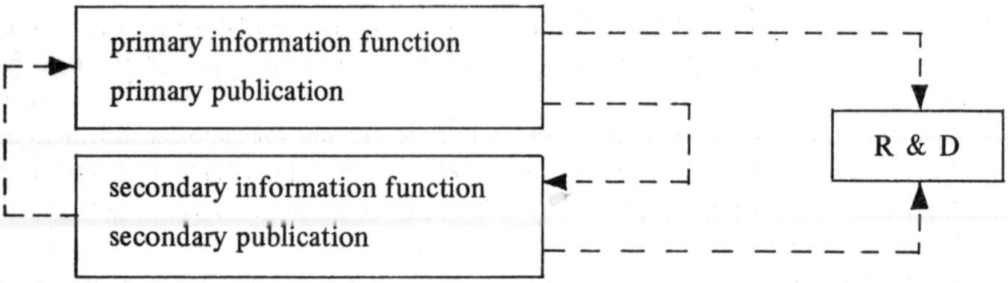

be the *minimum* of the "information officer's" processing work: apart from producing *subject-syntheses* it would not be concerned with the processing of primary literature (making abstracts, bibliographies, etc.). To put it another way, the work of the "information officer" *begins* where that of the documentalist *ends*. What is being considered here is a tertiary information function. This type of activity would be really an *organic part* of the research process, one of its most effective elements. It would demand, apart from a knowledge of the library and documentation, a high level of specialized knowledge of the *research area* in which he operates.

The establishment of the function of the "information officer" as described here, is at the same time a conscious formulation of the *tendency* that the users of documentation, of the secondary information, are in the first place *information specialists and not researchers*. If this is true, then this evaluation could be enlarged. It is not only relevant for the social sciences but also in a general sense.

The tertiary information function could not be done by any library or documentation centre, because their regular activities were oriented to the scientific community as a whole (in a branch of science) and not to *individual* researchers.

Implementing a new communication link, the tertiary information function, by means of *research units* (institutes, research teams) is one of the ways to make more useful secondary publications. It can be based on existing secondary services produced by libraries and documentation centres and partially based on primary literature. In other words it is necessary to *harmonize the contrast* between the tendency of the *scientific information apparatus to expand and the tendency towards a relatively decreasing use of it*.

Between the secondary and the tertiary information function there is not a rivalry but a partnership.

III. The equal partnership (between research and information) and the discrepancy between the demand for information specialists and their shortage owing to the greater attraction of research careers

It could be agreed, that in social sciences the literature search, the consultation of original texts, are *integral* parts of research work. The critical *evaluation* and *inter-*

pretation of special literature and information contained therein requires from the research worker a *more active participation* in the information process, than in science and technology.

The implementation of the tertiary information function in social science research could be considered as an important factor. The *indication* of relevant literature, data and facts preceded by the critical evaluation of the "information officer" puts the latter on an *equivalent level* with the researcher. In this case, there is an *equal partnership* between the researcher and the "information officer", who in fact *himself* is a researcher with *special tasks*.

But the research work constantly acts as an *absorbing* force attracting the specialists into traditionally recognized research work. Financial advantages, *prestige* and greater possibility of creative work create a *dilemma* for highly qualified specialists. A minor article in a scientific journal – and such articles appear by the thousands – qualifies as more than a very important subject-synthesis or a critical evaluation of the stage of research in different topics. The first one is considered "creative" work, the second one as "documentary" activity.

The higher intellectual and material *recognition* of certain information activities, the *re-evaluation* of frontiers between research and information in social sciences, the "equivalency" between research and certain type of information work, the *real integration* of the tertiary information function as *an organic part of research* – all these could stimulate and satisfy information needs in the social sciences much more than establishing new information centres and services.

In conclusion I would like to stress that the above points contain only a few basic ideas in the complex of the information needs in social science. They were formulated without a *definite project* in mind. They are *arguable* but the author thinks that the *formulation itself of certain points of discussion* could be helpful.

In: International Conference on Training for Information Work, Rome, 15–19 November, 1971. Rome, Italian National Information Institute, 1971. 315–320.p.

INSTITUTIONAL MATURITY AND SOME SOCIO-PROFESSIONAL ASPECTS OF AUTOMATED DOCUMENTATION

One of the most inciting new aspects that ameliorate the services on the ever growing number of written information (books, documents, periodicals, research reports, etc.) is automated documentation: one of the most important fields in the realm of information science. There are many achievements in this field, most in science and technology (and management), but they are also in social sciences (e.g. data banks). However, there is a gap in the application of automated documentation between these two great branches of knowledge.

And this is quite adequate. *Science and engineering* have, for the most part, an experimental character within which basic, applied and development research can be distinguished. Their information and data have an objective character, and the time factor (speed) plays an important role in acquisition of scientific information. In development research the results can be directly measured economically. The speed factor and the quantity of data requires the use of automated processing. As the *social sciences* have less of an experimental character, their results in the different branches again vary in immediate practical applicability (historical sciences and humanities, on the one hand, and concrete organizational and administrative branches, on the other). The information data have to bear an ideological aspect. In general, the demand for retrospective research is greater. The speed factor in acquiring data is not important. The information task in social sciences is, to a large extent, an integral part of the work of the researcher himself. The automated processing of information is not so pressing as in science and technology and requires some specific approaches.

Nonetheless no strict demarcation line can be drawn between these two large fields of science in their use of automated documentation. The distinctive features enumerated above afford a certain basis of differentiation.

Most of the studies on informatics in general and on automated documentation in particular, treat the question how to act, what are the problems of the techniques, programming, etc. In fact, the basic question of automated documentation is either what is to be done or why do it, and upon the answer of this question comes the know-how. It is not the automation technology which determines the concept of the automated documentation. On the contrary: from the concept the adequate technology or the technology "to measure" should derive. Under the notion of automated documentation, I mean a mechanized system which works through a chain of self-regulated procedures and without direct human intervention. At the same time it should serve and express the complexity of programs and services of libraries and documentation units, with special regard to the information storage and retrieval.

The power to do or the duty to do?

In the introduction of automated documentation, the circumstances of some factors such as the budget, the manpower, the computer, and others, available for a given project, very often play a preponderant role and the concept comes afterwards in conformity with the means provided[1]. Yet, it is less important what we can do; it is important what we *have* to do.

All this does not in any way leave in the shade the role played by techniques and technology. The mode of production is not defined so much by its products but rather by the level of the means of production and by the technology. The "product" of the automated documentation is of special nature: information. It is an intellectual product representing at the same time an economic value in some branches of social activities.

The accelerated development of science and technology, the needs of production as to the volume and the speed to receive new information, created conditions favourable to promote the information technology and its introduction in the libraries and documentation services[2].

The Marathon runner and thousands of congressists

One does not know how much time it took the Marathon runner to cover the 42 km which separated Athens from the battlefield to announce the victory. But we know that the death of Napoleon in 1821 at St. Helena was not known in Paris until 3 months later. Even in 1898, about 2 months passed after the events of Fachoda (Africa) and the reception of the first messages sent to Paris and to London by Marchand and Kitchener. A little more than a half century later, all the world (100s of millions of TV spectators of some 30 countries) directly viewed the first steps of man on the moon[3].

And what to say about the mass of information currently "consumed" by the 4,000 congress participants at the Congress of philosophy of Vienna (1968), 5,000 participants of the Congress of historians in Moscow (1970), and some those 10,000 invitees for the Congress of anthropologues and ethnologues in Moscow[4]?

Concerning the means of communication (the factor of speed) and the volume of information (the factor of quantity), these two examples have been cited expressly beyond science and technology.

These social activities should generate the technical means which could, potentially overpass the problems aroused by the volume of information, its treatment ("processing") and its diffusion. The computer occupies one of the foremost places among these technical means.

Priority: automated documentation or documentary management

Automated documentation is a field for implementation by computer. This field is characterized among others by the need for storing a great number of textual data information storage and retrieval and by the possibility of qualitative "retrieval" where the factor of speed does play practically no role.

From these statements we can conclude that the accent should be put on the preparatory intellectual work. The ensemble comprises at the same time questions of organization, professional training, and finding of the methods inherent to the automation such as the establishment of thesauri.

However, the realization of the automated documentation does not depend only on technical parameters but on the level of the management and of training. One could qualify this ensemble as institutional maturity.

The library or documentation or information management should therefore have priority in relation to the automated documentation. The objectives of the last are the arrival of the first mentioned.

The objectives of documentary management

Documentary management refers to the politics and to the general concept of an institution for scientific information. The management is the realization and the practical part of policies. It should be determined by the potential *sources* and by the *needs for information*. It should envisage the answers in a concrete way to the following questions:

– what is the role and what are the functions of the institution?

– what are its collections?

– who are the users?

– what kind of services does it offer?

– what is the size of its services?

– which are the internal services to be developed through traditional methods (including "small mechanization")?

– the same question relating to the external services whether they should be charged with the dissemination of information?

– what are the internal services to be developed through advanced techniques: computers, micro-reproduction?

– the same question relating to external services

– what can be the most appropriate structure to develop within the framework in which functions and tasks are executed?

In other words, first the general lines of a system analysis should be traced and afterwards this should be followed by the establishment of a feasibility study on automation.

Approach of a system analysis

An approach of a system analysis could be envisaged in the following way, in a kind of index, or a list of keywords for such analysis:
- definition of the policies of the institution
- what are its functions?
- analysis of the utilization in general and special analysis of these services
- the level of the management of documents (inventories, statistics, etc.)
- are there written instructions?
- methods and forms of written instructions
- social climate
- material conditions (budget, space)
- structure, organization, interrelation of units
- external relations
- plans and projects, short, medium, and long term
- professional training
- tasks of units, definition and description
- written instructions concerning different procedures (for example, treatment of documents, indexing of periodicals)
- flow of publications
- flow of information
- development of collections
- usage and facilities provided by its services
- methods of storage and conservation
- reproduction methods.

And so forth, bearing in mind the factors and elements most relevant to the functioning of the institution, one could determine the features formulated in the system analysis study which fixes the actual state of things and serves as a point of departure for future specialized analyses.

Approach to a feasibility study for automation

Based on the system analysis study, one could envisage feasibility study for automated projects. First of all, one should check which services could be improved through traditional methods. In other words, what are the factors and the elements of the existing documentary system to be developed which could be realized in a more rational and economic way without using expensive equipment.

Thus, one should envisage a *marketing* study with the aim of knowing what will be the services to provide, to whom they will be addressed, through what means and the implications of the cost-benefit. This marketing study, however, should not lose sight of the fact that information service has not only the task of answering requests, but also should incite, *stimulate*, initiate and even formulate potential requests. In this sense, the scientific information services accomplish also an educational role in attracting the attention of users to the possibilities unknown to them.

The feasibility study comprises also in the frame of the cost-benefit analysis, the repercussions of restructuring the institution and its methods in conformity with automation. The *cost-benefit* analysis should cover the general automated program even if this program is effected gradually. The culmination of such a program is the intellectual work and not the technology. The accent should be put on the documentary vocabulary and the descriptors, and on the thesauri which contain them, and on which the analysis of documents is made. This work is the most exigent. For this reason, special attention should be given to the possibilities of labor, even on international scale, as for instance the UNISIST projects. The co-operation and division of labour presuppose standardized methods.

All preceding statements could be converted into the notion of institutional maturity which is the "condition sine qua non" for any automation. No automated system, no computer without consideration of its capacity to memorize or its speed, *could put order into disorder*. The computer is an achievement for institutions and services which have *well defined policies*, are *well-organized* and work according to established *standards*. The computer is not an aim in itself as there is no universal remedy for all illnesses.

In many branches of social activities such as the research in social sciences, the information work is an integral part of the researcher, and the possibilities for automated documentation are more sophisticated. Institutional maturity means also that all these aspects are taken into consideration.

The farewell of the brave soldier Chwejk to sapper Vodicka

In the classical work of Jaroslav Hasek "The brave soldier Chwejk", there is among many others, one memorable episode in which Chwejk and his old friend the sapper Vodicka, after a quarrel with the soldiers of another battalion, were escorted to their respective units.

Chwejk invites Vodicka to meet at 6 p.m. in the "U Kaliha" tavern after the war in Prague. Vodicka accepts the invitation after being assured by his pal that "there is always something going on" at this tavern and asks him to come at 6:30 because it is very possible that he might be late. Everything is arranged for the rendez-vous: the place, the time (with a small foreseen delay), *except* the year in which the war will be over and this simple "detail": whether the two friends will survive the war. Therefore, everything was foreseen for this rendez-vous except the two principal factors: the year and the survival.

It is the same with information on the basis of automated documentation in the field of social information. One could have much information, useful details, even indispensable details, but these must be added to a concept or rather to a tool for preparation or for decision making, improving its realization. And that is the real *dimension* of automated documentation, especially in social questions.

NOTES

1. I leave aside other factors which could eventually play some role in the decision-making process concerning automation such as the "glamour of computers" the status symbol representing the machine, the interest of firms in "hard" and "soft"ware in the computer business, etc., etc.

2. See: A. I. Mikhailov–R. S. Gilyarevskyi: An introductory course on informatics/documentation. Paris. UNESCO, COM/WS/147, 208 p., based on the authors' "Osnovy informatiki": Moskva, 1968, Nauka, 766 pp.

3. Pierre Viasson-Ponté: L'information et l'action. *Le Monde*, 29–30 octobre 1972, p. 11.

4. La notion de culture et ses transformations (les délibérations de l'Académie des sciences morales et politiques.) *Le Monde*, 13 mars 1973, p. 14.

In: Study Committee FID/RI. Moscow, 24–26 April 1974. FID Publ. 530. Moscow, 1975. 355–363.p.

THE "WINNIE THE POOH DOUBLE-EFFECT"*

> Pooh complains that whereas the entries are wide the exits are narrow. He has to lose weight to get out. Unless primary literature "loses weight" we cannot successfully block the information tide.

Introductory remarks

The author's method and present treatment of his topic had caused a considerable controversy. The debate has not ended and the author has not finalised his position.

An approach to the problem

There are countless reports about "publication explosion" and "information revolution." The 1975 Boston Conference of the American Society of Information Science (ASIS) was devoted to the theme of "Information Revolution." The question is whether there is such a revolution in other words are we talking about a phenomenon or its main trends? The phenomenon is this: there is an exponential growth of total publications their excess is expressed statistically so no one can deny it. Pars pro toto: if there was a scientific-technological revolution then there is an information revolution too. And if there is a revolution on then we have to fight it with revolutionary means: with new technology, automation, computerisation.

But there can be another approach too. What is generally called an information revolution may just be considered as information inflation – growth beyond necessities. When we examine the terminology the content of information science – its mathematical problems, uses, needs, time-shares and ethical erosion – we can see that there is both a revolution and an inflation.

One may say that philosophy and theory apart we have to concentrate on real problems which means to see our way out of the information flood. This pragmatic approach is prevailing at the present time.

Could the "protection of the environment" – which tries to cope with pollution caused by overproduction – be regarded as some kind of revolution or simply a scientific-social recognition which aims to correct the anarchic elements of production?

---

* The author apologises for his title. It may just provide some diversion amidst reading the stern theories of information methodologies.

Not only nature is on the agenda but knowledge as well: its environmental protection, particularly in relation to scientific work. What is, therefore, the relationship between the information revolution and the "environmental protection of knowledge"?

The criteria of information functions

Narrowly defined, there are two meanings to scientific (library-bibliographic-documentation) information. First it signifies an *activity*, the organised transfer of secondary information. Secondly it denotes the *product* of the process which can be expressed in intellectual terms. Cybernetically speaking the product can be a sound, a sign, a datum and in some cases some significant piece of information such as a certain theoretical abstraction.

Scientific information transfers secondary data deriving from the primary in an up to date technical manner and form. It is the content of its up to dateness which is significant. This process gets rid of the irrelevant and concentrates on transferring the significant information,[1] i.e. the communication of instructive knowledge or the action of informing with some active and essential quality which is the area of primary communications.

The communicating of secondary informations, even in the case of their highest form such as "highlights" or "state of the arts reports" syntheses and studies by information centres cannot go beyond summaries of relevant informations on given topics so they do not communicate original information. Both activities are scientific work: the production of new knowledge and the information about it. But the first belongs to primary the second to secondary information. Yet research about secondary information and the theory consequent upon that belong to the first category.

The ammunition of information explosion

It follows from the above that even a partial solution to the information explosion cannot be either the developing secondary services or their international transfer; one has to take account of primary publications. There have been certain initiatives. In 1948 Professor Bernal, at a meeting of the Royal Society suggested the regulation of primary publishing[2] and academician Dubinyin in 1962 developed these ideas on regulating publishing.[3]

The significant features of these ideas relate to "synoptic periodicals" which replace the traditional ones while full texts would only be available in regional centres. The shock waves of the information explosion resulted in the formation of UNISIST which is the biggest international undertaking that deals with information. Apart from many useful programmes it promotes, its very existence is the result of the present inability to cope with the excess of primary publications.

There is justification for the working hypothesis that the growth of traditional communications (books, journals, lectures) and the non-traditional ones (reports and research surveys etc.) are not nearly in proportion with the quality and quantity of new information contained within them. This is not unknown to researchers. They say in conversations or in seminars that secondary information is useful only to a limited extent. They keep track of selected periodicals, they are interested in the range of books from which to select but often are indifferent or just occasionally interested in secondary information. Various sciences make out in various ways, chemistry, for instance is "information-hungry." The users of secondary information, in the main, are other scientific information centres. One should add, however, that the phenomenon of the restricted use of secondary material is manifested differently in different fields. While, for instance, the historian or the literary historian regards bibliography as their bread and butter a construction engineer is not interested in it. As for reference information or current awareness communications these are like adverts for washing powders (excuse the comparison) whose labels are different but whose contents are about the same.

A change of aspect

The argument above may seem negative and to a certain extent it is. What should scientific information services need, is there a model for them?

No one article can answer this question but one might attempt a list of experimental thoughts:

- there is no general model for scientific information;
- all information activity relies on the collections and services of a library – these have to be developed, above all;
- scientific information services are the best users of secondary information;
- in certain fields the role of secondary information in scientific development is in need of demystification;
- emphasis should be laid on direct syntheses such as "state of arts" reports;
- information scientists are members of research teams: their efforts to re-package information may seem important so long as the information will become immediately accessible;
- computers and information systems ought to be interlinkable;
- the basis of all computer based information systems is organisational maturity which includes personnel;
- the so called information explosion is limitable by a gradual and international reorganisation of primary publications.

About the double Winnie the Pooh effect

While the arguments of this article can hardly be summed up in mathematical formulae it may be paraphrased by two features in Milne's Winnie the Pooh. The two scenes seem to manifest the meaning of scientific information – the "double Winnie the Pooh effect."

The first effect relates to primary literature. Pooh visits Rabbit and gorges himself with milk and honey. When he wants to leave he cannot get through the door. He complains that while the entry is wide the exit is narrow. He has to lose weight to get out. One cannot hope for a radical change in the information-flood unless primary information loses weight.

The second effect relates to secondary literature. Pooh and Piglet go hunting. They believe they follow the tracks of a stoat whereas they only trace their own footsteps while going round and round. Christopher Robin tells them that in fact they have twice walked around the thicket.

NOTES

1. The Shorter Oxford English Dictionary on Historical Principles. 1973 ed.

2. J. Bernal: Provisional scheme for central distribution of scientific information. The Royal Society Scientific Information Conference Proceedings. London, 1948. pp. 253–258.

3. N. M. Dubinyin: Vazhnih voprosi obmena nauchnoi informatsie. Vestnik A. N. SSSR, 1962, No. 4, p. 40–43.

In: TMT, 1976.6. 239–242.p.

EFFICIENCY IN SCIENTIFIC INFORMATION, WITH SPECIAL REFERENCE TO THE SOCIAL SCIENCES

1. Are there parameters of efficiency in information work?

There are many publications concerning cost/benefit, effectiveness, and efficiency of information work. Mathematical methods and formulae are used to determine more precisely the meaning of these parameters.

It is a relatively clear what is meant by rentability. The rentability is an economic notion and may be applied to the measurement of certain aspects of information work. *The meaning of rentability* can be summarized as follows: *to obtain a maximum of performance with a minimum of input, having given implements*. The rentability in fact is an expression of the cost/economic effectiveness (or cost/benefit) relation. Its measurement is based on *comparative analysis* of different institutions with approximatively *the same functions, profiles* and *size*. It could be applied *directly* to information activity.

Efficiency, is a much more complex notion. Its measurement can not be expressed by economic-mathematical formulae. Efficiency does not relate only to economics, it is the expression of *a more general social relation*. Therefore, the application of quantitative methode for its measurement could give only partial results. *The evaluation* of the efficiency of information activities consequently *is more indirect*.

The rentability of the information work of a given institution could be high, but at the same time, its social efficiency could be low. What is good for a given institution and for its "clientele", is not necessarilly good on a social level. To give an example. Two institutions with a similar profile follow two different information policies. One of them disseminates up-to-date information, its publications (bibliographies, abstracts, etc.) are in high demand. The indexes of utilization of its files are high. In a word: the rentability of the institution can be considered satisfactory, and to some extent also its efficiency. The other is more oriented at searching for potential and new tendencies of research which are reflected in the special literature. This second institution collects materials for the future, which are not used immediately. The indexes of circulation of its information and those of the utilization of files are comparatively low. Consequently, its rentability is quite low. But at a certain moment, the development of its files oriented at new tendencies of research could become extremely important for the development of new research. It could even be said that the information activity of the given library of documentation center *has initiated and stimulated* a new outlook. I would consider that such activity is highly efficient. In fact, this is the meaning of the social efficiency of a given institution in spite of its lower rentability.

The question is, which one of the two institutions expresses better the meaning of efficiency?

The social efficiency of information activities *has no definite parameters*. As Marx said in his *Theories of surplus value* "The number of workers necessary to make a table, the amount of work of a definite kind necessary to manufacture a given product under certain production conditions are exactly known. This, however, is not the case with many 'non-material' products. The amount of work required to achieve a certain result is just as approximate as is the result itself". And if this is in general a true approach to intellectual work, to the 'non-material' production like research and especially social science research, it is also true for information activities, *whose results are even more approximate than those of research.*

A large amount of divergent publications could be quoted concerning the measurement of rentability and efficiency of research and development. The investigations are on an initial stage, especially those concerning efficiency. On the other hand, the justification of any scientific information activity is to provide information for research work. However, scientific information has its own features scope and methods, but it would be a kind of over-estimation of information work to study its efficiency disregarding its close connection with research. It seems to me that studies concerning efficiency of scientific information are in a stage of *defining the problems*. However, in this scientific approach, to formulate the points of discussions, the points of divergence could be considered as fundamental.

2. Development trend of scientific information activities: a brief statement

The development of scientific information following on the scientific-technological revolution imposes a new definition of certain activities that have evolved during the social division of labour comparable to the qualitatively new social function of science. Tentatively, the following can be said; in the process of science becoming a direct productive force, the communication of scientific knowledge, itself a product of this process, reacts upon it and becomes an *integral part* of it; one of its most essential and active elements. Like science, scientific information becomes organized, acquires mass proportions and tends to develop according to the organization of the productive forces.

This development, following a tendency, has several components, its analysis has economic, science-organizational, gnoseological and sociological aspects which require complex research. In the same way and by its very nature the process of transformation of science into a direct productive force shows itself most markedly in the natural sciences and engineering, so scientific information also acts as an *initiator* of the productive forces mainly in these fields. At the same time, however, the social sciences also undergo a significant development, especially in the field of *management* of society and economy, and in such branches as economics, sociology and management science.

The results of basic research in natural sciences, promoting both science and production, become applicable in their general and abstract character (e.g. mathematics, mathematical logic). These results gradually penetrate into social sciences, improving research possibilities and methods, *promoting* the development of new fields of research in the control, organization and registration of social-economic life (e.g. planometry).

The application of mathematics, mathematical logic, mathematical statistics and cybernetics, the introduction of concrete sampling methods and models, attempts at and experiments in quantification are becoming more and more characteristic also of the social sciences. In this connection, scientific information in the social sciences – to which it belongs – proceeds in the direction outlined above.

Thus scientific information has a two-fold significance: it is an integral part of certain areas of the social sciences which have a direct bearing on the development of the productive forces and it is a social-scientific activity affecting all the sciences. From this second aspect it follows that scientific information has to be approached as a problem of social sciences and according to *economic and science policy categories*.

The technological and scientific revolution gave a new impact to the development of the social sciences and consequently to the development of social-scientific information. This statement has a qualitative and a quantitative aspect, both being connected with the increase of specialized literature, documents and data, with information in general.

3. An approach to the efficiency of information work, with special regards to the social sciences

Generally speaking, it could be said that at least *two factors* of efficiency of information work are known. The first one is to have *well organized libraries*. They should be interconnected to form a *network*. This is the *basis* of all kind of information activities. The second one is the improvement of *research guides, information manuals*, the main problem of being informed today, is to know "how to find".

The acquisition of information in the social sciences is an *integral part* of the research. No bibliographic, abstracting or documentation service can replace the information investigation made by the social researcher himself. These services could provide extremely valuable assistance but their role is not to be compared with similar services in natural sciences and engineering (e.g. Chemical Abstracts or Chemical Titles in chemistry). Many information services are doing their best for social science research on the one hand, and important reference literature, like research methodologies, bibliographical surveys, introductory manuals in various branches of social sciences are given some orientation to research, on the other hand.

What is to be developed are introductory information manuals for research ("information propaedeutics") for different branches of the social sciences. By information manuals, we understand *a special type of reference manual* combining some aspects of the "traditional" reference works (e.g. bibliographies, list of periodicals) and those of research methodologies.

The main task of this new type of information manual is, to provide a general and *systematical survey* of a certain branch of social science, an overview on its institutions (research apparatus), sources of references, library and documentation services and periodical literature. Each of these chapters contains *the state of the art*, an analysis and a *relevant bibliography*. The state of the art are based on *typology*. By typology we mean not only registration and classification of relevant material on the source of references, on documentation services, etc., but analysis of all these from the point of view of their *application in research*.

The manuals should concentrate on *summarizing, describing, systematizing* and *characterizing* all factors and elements effecting social science research (by branches), covering their research methods and institutions (scientific and practical organizations), basic periodical literature, information sources (bibliographical sources, library services, encyclopaedic publications, etc.), chronologies, statistical compilations, computer based systems, that are necessary in the given field of research.

These information manuals ("propaedeutics") could give an orientation to individual researches, to research teams and also to reference institutions. They could be used in higher education and training *to find the types* of institutions, of reference materials, in a term, what is the *infrastructure* of a given study.

In: On effectiveness of scientific information activities. Study Committee FID/RI. FID Publ. 527. Moscow, 1976. 36–40.p.

SOCIAL SCIENCES INFORMATION: TYPOLOGY OF SOURCES

Based on the example of economic research, the author outlines a method which permits the drawing up of a typology of social science information sources.[1]

Social sciences like the natural sciences are not homogeneous in character, and like the latter comprise several disciplines and professional activities related to a wide range of study fields. While taking into account some general aspects of the social sciences the heterogeneity of these sciences implies that research in each of the social science disciplines requires its own, specific information.

It should be noted that the meaning of the "social sciences" has several interpretations. In general in the non-Marxist terminological framework, social sciences do not include the humanities, while from a Marxist point of view social sciences comprise the humanities such as literature, linguistics as well as history.

This article attempts to formulate a common basis that could reflect both interpretations of the social sciences. Such a general basis could be arrived at by tracing and systematizing the sources of information for social science research. The typology developed here is an "information-oriented" one and it could be applied to research in international relations, economics, law, sociology, cultural anthropology, etc. To exemplify the application of this typology I have chosen from the social sciences the field of economics because as a social science branch it is many-faceted, and as a discipline it is as an amalgamation of several branches of knowledge.

However, before discussing the typology of economic information sources, a brief comment should be made on economics itself. One of the characteristic features of economics is its closest connection with practice, with theoretical and practical work highly intertwined. Information in the field of economics stems partly from the traditional organizational frameworks of research (universities, research institutes) and from a variety of research sources, e.g. conventional monographs and periodicals on the one hand and a multitude and variety of business reports, market news and analyses, bank reports, statistics, departmental bulletins, all concerned with the daily, practical problems of economic life on the other. If we add to all this the press with its daily and weekly newspapers, an extraordinary variety of economic information sources is further compounded, especially if information from related disciplines such as sociology, economic history, management, etc., are taken into account.

The aforementioned variety of economic information sources is very suitable to illustrate our typology. The sources of research in economics may be grouped into two large categories such as "academic" and "non-academic" categories. The former consists of literature which emanates from traditional research and professional organizations while the latter pertain to economic almanacs, market research reports, etc., i.e. publications of various economic-business organizations. These two categories are organically connected. A set of uniform principles permits the systematization of the sources within these categories. The main criterion of such typology is that the systematization of economic information should reflect specific aspects of economic research (theoretical and practical research, grouping of research by branches). In this way organizational (institutions) and formal (i.e. forms of special literature) criteria may be linked with disciplinary (in this particular case with the economic) aspects. Hence it follows that only the "main lines" of the typology may be general (e.g. the linking up of academic and non-academic categories, or of organizational disciplinary aspects), while the typology of the individual social science branches should be carried out separately in accordance with the specific features of a given field of research.

The "general" aspect of the typology, therefore, is the "method":[2] the harmonization of organizational, formal and information aspects with the specific aspects of research. With all this in mind, let us take a look at the aforementioned method as it is applied to economics.

What sort of information, organizational and formal, is needed by researchers and professionals in economics or by information specialists? The first large information category relates to the "research apparatus" (Synoptic Table A) in terms of the

SYNOPTIC TABLE A. Systematization of the "research apparatus" of economics

I. By area	International	Intergovernmental
	International regional	Non-governmental
	National (by countries)	
	National-regional	
	Research organizations (learned societies; higher education; research institutes)	General (e.g. academies of sciences)
		General (social sciences)
		General (economics)
		Economics by branches
		Economics-adjacent (limitrophe) (e.g. economic history, economic law)
		Others concerned with economics (e.g. agricultural, industrial)
	Official organizations	International (general; economic)
		By national level (e.g. planning offices)
		By regional level
	Statistical organizations	Offices (international; by national level; by regional level)
		Associations (international-general, by national level; by regional level; by branches)
		Higher education
		Profit organizations (e.g. Banks)

SYNOPTIC TABLE A. *(continued)*

II. By type	Social organizations and societies (non-profit organizations, trade unions, foundations, etc.)	
	Business organizations	Business federations Banks, financial institutions Firms, multinationals Marketing institutions Public relations
	Managerial (profit) organizations	General By branches Productivity agencies
	Information organizations (see Table D)	Adjacent (limitrophe) organizations (e.g. labour psychology) Management consultants

SYNOPTIC TABLE B. Systematization of the "reference sources" of economics

By type	General scientific publications, propaedeutics	Encyclopedic works Propaedeutics Manuals, basic textbooks Yearbooks and serials Research methodologies Directories Bibliographies
	Collections of materials and data	Official publications, documents (e.g. national plans, budgets) Official publications, documents of social bodies General economic descriptions, manuals Reference yearbooks and manuals (e.g. *The Statesman's Yearbook*) Chronologies (e.g. *Keesing's Contemporary Archives*) Statistics
	Aids of a technical character	Commodity and price lists, currency lists Business directories Technical lists (e.g. on standards, on licences) Language guides Topographic guides Communication and transport guides Fair and exhibition guides Public relations materials

organizational framework (learned societies, institutes, companies, marketing agencies, and so on). The second category (Synoptic Table B) is connected with the "reference sources" of economics. The third category (Synoptic Table C) pertains to "periodicals", while the fourth (Synoptic Table D) comprises "information services" (including libraries, data banks, documentation agencies) in the field of economics.

The typology of the information research sources may be drawn up according to various principles. For example, the systematization of the research apparatus is carried out mainly in accordance with the "territorial principle". Thus we may speak of international, international regional, national research institutions and or regional within a certain country. The grouping by international and national research organizations may have a double meaning or interpretation. A research organization may be an international institution as, for example, the International Institute of Statistics (general

SYNOPTIC TABLE C. Systematization of "periodicals" in the field of economics

I. By area	International
	By national level
	By regional level
II. By periodicity	Annual (yearbook)
	Journal (monthly, bimonthly, etc.)
	Weekly and daily press
	Serial and other non-regular published material (e.g. newsletter)
	Press releases (irregular)
	Press agency news (several times per day)
	Business communications (several...) (e.g. stock exchange)
III. By publishing (including intellectual, or corporative publisher)	Scientific body
	Official publication
	Publishing house
IV. By type	Business publication (e.g. bank reports)
	General journals
	General social science journals
	General journals on economics
	Journals by branches of economics (applied economics)
	Economics-adjacent journals
	Official publications
	Statistical publications
	Business publications (e.g. bank reports)
	Management journals
	Weekly and daily press (economics colums and news): General Specialized in economics
	Information publications (see table D)

SYNOPTIC TABLE D. Systematization of "information services and supplies" in the field of economics

I. Information services	Libraries	General libraries Special libraries	Economics in general By economic fields
	Documentation organizations	Social siences Economics	
	Data banks	Social sciences Economics	
II. Information supplies	Reading rooms		
	Loan services (including interlibrary loan)		
	Catalogues		
	Reference services (reading lists, literature search)		
	EDP-services (SDI-service, data-bank services)		
	Specialized supplies (e.g. press clippings)		
	Special collections (manuscripts; archival materials; maps)		
	Audio-visual materials (records, photographs)		
	Reproduction supplies (xerox copy, microreproductions)		
III. Information publication	National bibliographies		
	Special bibliographies		
	Abstracting and indexing services		
	Union catalogues		
	Printed (library) catalogues		
	Acquisitions lists		
	Lists of translations		

international) or the United Nations Economic Commission for Europe (international regional). The aforementioned two institutions are international in terms of both their organization and their scope of activities, fields of study. There are, however, institutions which are "international" only in term of their fields of study, the Royal Institute of International Affairs (London) or the Institute of World Economics and International Relations of the Soviet Academy of Sciences[3] being examples of this type. Moreover, international institutions may be intergovernmental or non-governmental organizations. The grouping by countries or national organizations may also have a double meaning; the work of a given research organization may reflect exclusively national economic activities, or its work may explicitly be focused on the economy of a country other than its own. These area considerations essentially apply to the grouping both of monographs and periodicals.

The most characteristic grouping, however, is the systematization by type. Research organizations can be said to belong to the following main types: academic institutions; official organizations; statistical organizations; social organizations and societies; business, management or information organizations. The three main groups of academic research organizations are: learned societies, higher educational institutions and research institutes. Research organizations, by field of activity, may be science in general (e.g. National Science Foundation, United States), social science in general (e.g. Maison des Sciences de l'Homme, France), economics in general (e.g. Institute of Economics of the Hungarian Academy of Sciences); branch specific within an economic field (study of finance, foreign trade); interdisciplinary with an economic aspect (economic geography) or research organizations in fields other than economics (e.g. technological, agricultural research organizations).

In addition to the aforementioned academic research organizations (ministries and other high government bodies), statistical organizations, social organizations and organs (e.g. political parties, foundations, trade unions), business and managerial organizations, as well as economic information services, all play a considerable role in economic research.

There is no strict dividing line between the academic and non-academic research organizations; in many cases their activities overlap and they frequently use one another's output. Therefore, from the viewpoint of a given research project, information service from both categories should be taken into account.

The academic and non-academic aspects based on uniform principles also apply to other categories in the typology of economic information sources, i.e. to the reference sources of economics, to periodicals and information services in the field of economics. Thus, economic information sources will include among other things, commodity and price lists, business directories, periodicals, press cuttings from daily and weekly newspapers, etc. In general, it would seem that the systematization of the research apparatus by *type* can be applied to different categories in the typology of economic information sources (cf. Synoptic Tables A. II. and C. IV.).

In its present general form and through the example of economic research, this article attempts to contribute to ongoing endeavours[4] in the field of the systematization, classification, etc., of social science information. For this method to be applicable to other areas of social science research the typology as outlined above should be developed by providing greater details, and the systematization of the individual categories should be further improved.

NOTES

1. A preliminary elaboration of this topic can be found in the following studies of the author: *Some Ideas* on a *UNISIST Sub-system for Social Sciences*, Unesco, 10 September 1973, 12 p. (mimeo.); "Efficiency of Scientific Information with Special Reference to Social Sciences", *On the Effectiveness of Scientific Information Activities*, p. 36–46, Moscow, VINITI, 1976. (FID/RI Series of Collected Articles, 527.)

2. Cf. *Typologia Dokumentow; referaty II. Miedzynarodowego Spotkania Ekspertów Krajów Socjalistyczny d/s Biezajec Bibliografii Narodowej. Warszawa, 21–26 kwiecieri 1975.* Warszawa, 1976. 104 p. (Biblioteka Narodowa, Prace Instytutu Bibliograficznego, 21.)

3. Institut Mirovoj Ekonomiki i Mezdunarodnih Otnosenij, AN SSSR.

4. Mention should be made of the classification systems used in the series "International Bibliography of the Social Sciences", London, Tavistock, particularly in its *International Bibliography of Economics*, Vol. 1, 1955–.

In: Unesco Bulletin for Libraries, 1978.3. 167–171.p.

INTERDISCIPLINARY RESEARCHES AND INFORMATION SYNTHESES

Informatics may essentially be interpreted in two ways. Firstly, informatics may be understood as a highly comprehensive and complex concept of the theory and practice of scientific information. This embraces the totality of activities involving library, documentation and the automation of information. Secondly, informatics may be regarded as a general application of cybernetics. As understood in the first case, information syntheses are ranking among the new trends of informatics. These have been strongly developing over the past fifteen years. This development has been influenced considerably by the information needs of interdisciplinary researches. At the same time, this type of reserach work expresses the integrative tendency as against the high degree of specialization of research and development.

With some exaggeration, though, it may well be stated that a good part of the really effective contemporary research efforts is interdisciplinary in character. It is a new phenomenon that the interdisciplinary trend as it appears *within* the large scientific spheres – sciences, social science and the humanities – will increasingly prevail *among* these spheres. The individual disciplines of these spheres will intersect, will melt in each other. To be mentioned among scores of instances is the protection of natural environment that involves – among others – biology, geography, medicine, technology, politics, economics, sociology, social-psychology, or prognostics as an integrated social science research that adopts mathematical methods, too, but to be quoted as simpler example are mathematical linguistics, ergonomy, and scientometrics.

It is by far no news that there is a partnership between literary science and linguistics since they have always used each other's achievements, their research works have been intertwined. Connected with the development of sociology, there appeared the sociology of literature, as well. However, these "intersections" have been seen only within the sphere of social sciences and humanities. Different is the situation in the case of mathematical linguistics, computerized translation or lexicography which cannot be imagined without the involvement of the sciences. Generally, the spread of the large-scale adoption of mathematical methods in scientific spheres other than natural sciences have had a strong effect on the development of interdisciplinary researches.

But there are also certain branches of science that have not become, but were precisely *born* to be interdisciplinary studies. In addition to the above-mentioned environment protection, such study is space research which came into being by the agency of mathematics, astronomy, meteorology, biology, medicine, nutrition (or dietetics), psychology, telecommunication, metallurgy and of a wide range of other disciplines that cannot even be enumerated.

Sándor Szalai, a sociologist, member of the Hungarian Academy of Sciences thus formulated these phenomena as the main types of interdisciplinary researches.[1] There are *small-span* researches in which neighbouring, closely connected branches of science and scholarship are participating such as legal science, economics, and so forth. Then there are *medium-span* interdisciplinary researches such as those connected with public health or urbanization. Finally there are *immensely broad-span* researches like the above-mentioned space research, which call for the participation of disciplines which until now have been almost completely separate, having quite different conceptual, organizational and institutional systems, and have had hardly any intercommunication. Such are, for instance, astronomy and psychology, optics and microbiology, and so on.

Formerly, until about the Second World War, one could see the development of *borderline sciences*. However, the recent development can no longer be characterized simply by borderline researches (e.g. biology and chemistry, which gave birth to an independent new branch: biochemistry), much rather can it be characterized by interdisciplinary researches, not caring a fig for the traditional classification or systematization. "Interdisciplinarity" itself is a process of integrative character. And all these have their own impact on information.

The process of change in the scientific approach that is going on in research and development in the direction of the integration of knowledge, should make its effect felt on information too.

From the aspect of *orientation*, the following main types of information may be discerned in scientific information:

Information system oriented by *the kinds of documents* as for instance, the information system of non-published documents of on-going research, or patent documentation, or a UNDEX, the bibliographical control system of UN documents.

The *discipline-oriented* information system is the most wide-spread. Practically every branch of research and development has its own traditional and/or automated system (for instance Referativni Zhurnal of the VINITI).

Mission-oriented information systems are essentially the systems of such interdisciplinary researches or projects as space research, environment protection, development science (problems of the developing countries), science of science, or informatics that cannot be classed in one discipline.[2]

Presumably, we will be or rather we already are witnessing a developmental tendency in the course of which alongside with the mentioned various "oriented" information systems there will also be an information service which is characterized by the synthesis or a high-level, expert summary of the given subject field. Can these services be termed as being *synthesis-oriented*, or rather as *research based information*? But it is not the term but the *tendency* that really counts.

What does this tendency express? On the one hand, information synthesis is a response to the fragmentation caused by the specialization of information, sort of a scientific reply, a *quasiencyclopedic headword* concerning a certain scientific problem or theme. In other words, it tends to express the *integrative tendency* of information that is adequate to the interdisciplinary trend in research and development. On the other hand, information synthesis is one element of *decision-making information*, particularly in the preparatory phase. These two basic lines of information synthesis, however, does not exclude the possibility of other variants. To be found among these further variants are the summaries of a well-defined problem field of a discipline for research purposes or the concise description of a certain topical question for practical purposes such as – for example – information works of the Library of Congress Congressional Research Service for members of the Congress.

What is common in information syntheses? The first and foremost common feature is that they supply textual, sometimes factual *evaluated* information mostly by the processing of *several sources*. This is practically done by the utilization of any source material that is related to the given topic independently of disciplines, regarding only the value of the source and its relevance as the criteria for selection. These criteria are adequate to the requirement of *selectivity* and *complexity*. The *comparative approach* may also be mentioned as a characteristic feature of the work of selection and processing. The simultaneous presence of these criteria permits to give a high-level survey over the given subject that may substitute for the reading of the original primary literature, or in the case of a large-scale research work it may introduce the user in the subject matter, and it may be the starting point of research. The first phase, then, is the selection of the literature and data, their *scientific analysis*, and this is followed by the synthesis. This is why the special literature refers to this kind of information as *analytic-synthetic review study*. The name of this kind of information is highly varied: *the state-of-the-art report, itogi nauki* (progress in science), or surveys prepared by *information analysis centers* are all covering identical concepts.

In the above enumeration there is one conspicuous thing. Beside information synthesis and its synonyms, related terms which all denote products of information, there appears an organizational frame-work, the institutionalization of information syntheses. In fact, information analysis centers, which came into being in the sixties, were set up to prepare information syntheses in a regular, professional or let us say "industry-like" manner. As regards their character, these centers are ranking somewhere *between* research institutes and information institutions, in the same way as information synthesis takes an intermediary place between information and research report. As to which of them in information synthesis is nearer (considering both the organization and the product) to information work or to research, can hardly be defined with universal validity. However, what can be stated is that information synthesis is by all means a type of information *standing nearest* to research report.

Beside acting as a transmission agent between specialized literature and research work, this type of information may become a specific transition between documentation and research activity. To use an analogy taken from industry, its task is not simply to provide research with material, to meet its demand for raw materials, but also to turn out "semi-finished research products". By *"semi-finished research products"* we mean the processing of documents into tabulation, the rendering of the data methodologically commensurable, the evaluation of their proportions, the analysis of documents (organizational and administrative schemes, science budgetary systems, computation methods for assessing economic efficiency, etc.), and their international collation. By fulfilling these tasks the documentation of science organization (to which we shall revert later) also performs tasks of a limited research character.[3] Therefore, this form of information may also be termed as *tertiary* information supply. On the other hand, information analysis centers are standing near to research institutes.

Which are the criteria for these information centers? The consideration of these criteria will cast more light on the characteristics of information syntheses. The information working group of OECD thus outlines the operational principle of information analysis centers:

– It works in a well-defined subject field to serve an adequate number of users;

– works under scientific leadership;

– is closely connected with the leading researchers;

– 75 per cent of the received inquiries will be replied by an expert of the given subject field;

– at least 50 per cent of the information supplied goes beyond bibliographical information;

– it issues data collections and analytic-synthetic reviews.

By and large, the same operational principles are deemed as characteristic by the authors of the basic work on informatics.[4]

However, the information analysis center is nor the only workshop of publishing information syntheses. Information syntheses may be just as well issued whether by a research institute or by a library or documentation center. The Library of the Hungarian Academy of Sciences is a case in point.

This library has been publishing information syntheses in a regular way ever since 1961 in the field of science of science, in the form of a documentary periodical titled Bulletin of Science Organization (Tudományszervezési Tájékoztató)[5]. This periodical was called forth by the actual needs of Hungarian science policy, at a time when the problems of the top management, financing and planning of the country's research effort had come to the fore. Its backbone, its most important part is the column of information syntheses. One issue of the periodical contains 8 to 10 such syntheses (originally termed as "review articles") mostly by external specialists. The editorial staff is not engaged in library work in the proper sense. Their task is regularly to keep track of several hundreds of periodicals which contain – with more or less regularity – articles, studies on science of science, and also to examine the new book acquisitions of the library from the aspect of science of science. Their task is also to be in touch with the users (research managers, research institutions) and to recognize the subject fields where the need for information syntheses arises, comparing the potential needs with the literature available. Not infrequently do users give ideas and proposals to the editors. In parallel with the formulation of the plan of the individual issues, the designation of the most suitable authors is also carried out.

Let us take the number 6. of the Bulletin's 17th volume (1977) to exemplify the thematic fields covered: "Possibilities of assessing the development of industrial research units and certain problems of measurement" (by a senior scientific researcher of the Research Institute for Telecommunication) "Can success cause trouble?" (by a staff member of the Economic Information Group of the Hungarian Academy of Sciences) "The present state and future prospects of the Swedish Social Science Research Council" (by an editorial staff member) "The history of a large institute of the Hungarian Academy of Sciences" (Institute of Automation and Computing, by the deputy scientific secretary of the Institute) "The scientist's personality" (by a professional translater) "Research careers and career structures" (by a psychologist) "Tsukuba Japan's new Ideopolis" (by an economic researcher) "Preconditions of creative cooperation in university research" (by a senior scientific researcher of the Polytechnical University). Let me give you some references to the last review article (indicating only the title of the periodical and its place of publication): Le Monde (Paris), Das Hochschulwesen (Berlin, GDR), Research Management (New York), Arbeit und Arbeitsrecht (Berlin, GDR), Impact of Science on Society (UNESCO, Paris), Sovremennaya Vizshaya Skola (Warsaw), M. Beck: Tudomány – áltudomány (Science – pseudo-science), Budapest, Rinascita (Rome), Spektrum (Berlin, GDR), Pravda (Moscow), Előre (Bucharest), Philos (London).

The contents list, the authors of this issue and the references speak for themselves. The information synthesis column of this periodical is *one variant* of this information type. The articles are dealing with topical problems, using selected and evaluated source materials, their elaboration is complex, comparatively concise, contain rich information, and the author is an expert in the given subject. Finally some data: in 1977, the review articles processed 280 documents. In six issues a year, counting 8 to 10 review articles in each issue, an average of 4 to 5 source material were processed in one article.

Science of science, moreover, is one of those disciplines of integrative character that came into being on interdisciplinary basis.

Like any other scientific special literature information, information synthesis is also based on the document collection. In this context, the carriers of "interdisciplinarity", the integrative tendency of scientific knowledge are *the large academic libraries*.

As against specialized libraries these large academic institutions reach beyond the information supply of ongoing research towards whatever appears as novelty in international science. And they do this not out of indulgence but as a matter of duty. One of the most important tasks of large academic libraries lies in the collection of borderline and interdisciplinary scientific literature or the collecting for those new tendencies which have not yet been organized into a discipline. In this way they can give effective help to research, and can become a basis for interdisciplinary research and information syntheses. These libraries are built on a heritage of Diderot and D'Alembert, the strivings of the 18th century French encyclopedists after the integration of knowledge

In his Colombus biography, published in 1940, the Spanish philosopher Madariaga put forward the hypothesis that the discovery of America was due to the fact that in the Royal Library of Lisbon Colombus had secretly made a copy of the letter and map of "the western route" of the Italian astronomer Toscanelli. The Colombuses of contemporary science should no longer have to resort to the secret copying of documents in libraries. Immense scientific library and information organizations are working in their favour, bringing research and information in ever closer connection. And this is the vital part of informatics.

NOTES

1. Introductory paper (manuscript) by Sándor Szalai at the MISON conference on the information supply of interdisciplinary social science research (April 11–13, 1978, Budapest, organized by the Library of the Hungarian Academy of Sciences). MISON is the Russian acronym for International Social Science Information System which was set up by the academies of seven socialist countries (Bulgaria, Czechoslovakia, GDR, Hungary, Mongolia, Poland, USSR) in June, 1976.

2. UNEP/IRS (United Nations Environment Programme (International Referral Service), ESRO/ /RECON (European Space Research Organization, uniting the western countries), DEVSIS (Development Science Information System), a system initiated by the Canadian International Development Research Center, CORE/IOB (Common Register of Development Activities/UNDP Inter-Organization Board of Information and Related Activities), SPINES (Science Policy Information Exchange System), initiated by UNESCO, ISORID form the international information system on informatics (UNESCO/FID).

3. György Rózsa: Scientific information and society. The Hague – Paris, 1973. Mouton. 159 p. cf. 101–202. p.

4. On the basis of a study "Analytic-synthetic review studies" (in manuscript) prepared for a textbook by Tamás Földi, director of the Economic Information Group (sort of an information analysis center) of the Hungarian Academy of Sciences: OECD. Scientific and Technical Information Group. Misconfrontation on information analyses centres. Paris, 1970. and Mikhailov, A.I., Cernyi, A.I., Giljarevskij, R.S.: Nauchnie kommunikatsii i informatika. Moskva, 1978. Nauka. 536 p.

5. The current volume of the periodical is the 27th in 1987. It appears in six issues annually in 900 copies. Some 120 copies are distributed abroad on subscription or exchange basis. Its columns are: Reviews (information syntheses), New and Views (minor articles on a certain subject on the basis of one document), Bibliography (partly annotated and regular bibliographies of the special literature, international and Hungarian), contents and summaries of reviews in English and Russian. Since 1983 the title of the journal has been changed into *Kutatás-Fejlesztés (Research and Development)* as its subtitle remained *Bulletin of Science Organization*.

In: New trends in informatics and its terminology. Study Committee FID/RI. FID Publ. 568. Moscow, 1979. 103–110.p.

MAKING USE OF INTERDISCIPLINARY INFORMATION: THE HUNGARIAN ACADEMY OF SCIENCES

The Hungarian Academy of Sciences, established in 1825, is the country's highest scientific authority. It acts as scientific adviser to the government, operates one of its most significant network of 45 research institutes from virtually all branches of science, and also finances and controls a number of researches at universities. Its information centre is the Library of the Hungarian Academy of Sciences, one of the most important scientific information bases of the country. Both the Academy and its Library are the creator, organizer and disseminator of interdisciplinary information.

1. Historical background

The Library of the Hungarian Academy of Sciences was set up in 1826, only a year after the establishment of the Academy itself. Preceding the II. World War it has already become one of the three most important libraries of the country, and from 1948, together with the library-information units of the Academy's network of research institutes (comprising 45 institutions) it became the largest scientific information base in Hungary with the widest spectrum: it owned 2.5 million bibliographic units and 9 thousand current periodicals representing almost all branches of science.

The information network of the Hungarian Academy of Sciences is very wide in scope and has many aspects. This variety reflects on the one hand the wide spectrum of the Academy's research work, on the other, the degree of integration in information evolved during the course of historical development. This integration consists of the unity of the library – archives – automated information and its interdisciplinary character represented by the Library of the Hungarian Academy of Sciences. This Library is the center of the information network of the HAS.

The Library of the HAS has been developed into an information centre in its widest sense. Into this centre have been chronologically integrated with the library the Archives of the Academy and the Natural Sciences Information Directorate, established later, and being engaged in computerized SDI and science analysis (scientometrics). As part of the process of continuous broadening of library tasks, preliminary studies are being conducted with international collaboration to build up a computerized data base for the social sciences.

2. Information integration and interdisciplinary needs

The basis of interdisciplinary information services, besides comprehensive collections, is the integrated library – archives – informatics service, realized by the Library of the HAS. This interdisciplinary information is the joint result of the wide scope of the contents of the holdings, the types of the collections and the combination of the functions and services. Information services range from conventional reference work to the use of computerized data bases – expressed in UDC language: from the 0 to the 9 class.

The creation, organization and dissemination of interdisciplinary information at the Hungarian Academy of Sciences and its library has historical and, in its application to research, practical reasons.

The historical cause is simple. Until the country's Liberation in 1945, the HAS, set up principally with the help of donations for the purpose of promoting the Hungarian language, its library maintained by donations and international exchange of publications, had been an exclusive establishment with a bias to the humanities. It is worth mentioning though that the periodicals collection of the library had already been quite important in the sciences as well, as a result of the international exchange of publications (Proceedings, Compte Rendu's, Abhandlung's of foreign Academies). Furthermore, the Library had been the sole establishment of the Academy.

The practical causes are more varied. Following the reorganization of the Academy after 1948, numerous specialized research institutions sprang up from the field of science, social science and the humanities. These institutes have been carrying out basic research and each developed its more or less significant scientific information service. Some have achieved national importance, like the Biological Research Centre (Szeged), the ATOMKI (Institute of Nuclear Research, Debrecen), the Central Research Institute of Physics (Budapest). With the professional and methodological guidance of the central library, these information centres formed into a network. Meanwhile the functional system of the central library has also undergone a change: it had to expand to qualify for the provision of a dynamic information service both for the Academy and for Hungarian scientific life as a whole. Among other things it had to be gradually suitable for:

a. co-ordinating the libraries and information centres of these research institutes (meaning to some extent co-ordination of their acquisitions and exchange of publications, etc.);

b. representing the Academy at international professional organizations (such as MISON, ECSSID – the ECSSID Bulletin is being published here – etc.);

c. the keeping of central academic and national scientific records (e.g. compilation of the bibliography recording scientific works of the numbers of the Academy, the collection of candidatural and Academic doctoral theses, registration of travel reports from abroad of Hungarian scholars);

d. provision of central scientific information services, such as science of science, research management, science policy (23 volumes of its journal have been published so far);

e. publication activities based on the library's own collections with the cooperation of its personnel (series such as "Publications", "Informatics and Science Analysis", catalogues of the Department of Manuscripts);

f. the collection of non-book and non-periodical type material such as the Archives of the Academy operating as an independent department with its archive-, tape- and photographic collections;

g. the introduction of up-to-date information technologies, such as micrography, offset reprography, xerography and alike;

h. and last but not least the modernisation of information techniques, which started up a few years ago with the SDI services based on the magnetic tapes of the Science Citation Index.

All these, and others not mentioned, are user needs centred or user oriented services, manifested functionally as integration, and in contents as interdisciplinarity (a question of terminology, as the phenomenon could equally be called multi- or pluri--disciplinarity). The items under a–h. have a common characteristic: all reflect in a more or less direct form the institutional or personal research need.

3. Some characteristics of interdisciplinary information

As I mentioned before, the HAS and its Library (or through its Library) is the creator ("generator"), organizer and disseminator of interdisciplinary information. Data specified in the previous chapter outline chiefly the research-user requirements represented by the Academy, and the items of the programme aimed at satisfying these requirements by the library. But we have to proceed further than that.

The Academy does not only represent interdisciplinary interests but is itself an interdisciplinary institution (its present 10 scientific departments embrace all disciplines) and the majority of the results of researches conducted by its network of institutions is being published by its own publishing house, the Publishing House of the HAS in its periodicals, series and monographic studies. It issues over a hundred periodicals, approximately half of these in foreign languages ("Acta"-s), and partly in cooperation with major international publishers (Elsevier, Pergamon, etc.). A number of these serials are distinctly interdisciplinary in character, as for example "Scientometrics" (its editorial office is in the Library of the Academy, its publishers are Elsevier and the Publishing House of the HAS).

The Library of the Academy, like the Academy itself, also represents, or rather accomplishes this interdisciplinarity through its structure, collections and services (see 2. chapter, items a–h.). The Library is presently an information establishment which is integrated in its organization, interdisciplinary in its character, and combines the functions of the conventional library, archives and computerized information as well. It is a general library (tradition retains its name "library" without indicating an information institution) and also a national special library in some fields, such as linguistics and literature, classical studies, oriental studies and "science of science", with very important periodicals collections in mathematics, physics and natural science. The composition of the users of the library reflect its interdisciplinarity.

Which are the features that can be generalized, or rather, that are general, taken from the interdisciplinary establishment or information service detailed above?

a. The trend toward specialization which began in the 19th, and intensified during the first half of the 20th century, points, as a consequence of scientific development, in the direction of integration and interdisciplinarity. This course is being followed by the information service, or rather, by way of its publications and secondary services the process is being encouraged, and in turn greatly advanced by the possibilities of computerized information.

b. The interdisciplinary information services observe world science in general (what new advances in research?) they induce new researches, while the specialized services satisfy current researches and concrete user requirements.

c. While reference services of libraries (particularly those that acquired a greater role in the 19th century) followed the requirements of researcher-users, the information services in the latter decades of the 20th century have become initiators. Their work has become more and more incorporated into the research itself, thus assuming a more creative role.

* * * *

There is a connection between interdisciplinarity and creativity and not all of its aspects have been analysed so far. I have attempted to show what can be done in terms of certain information services.

In: The use of information in a changing world. Ed. by A. Van Der Laan and A. A. Winters. The Hague, 1984. 171–174.p.

"TO CONNECT THE COUNTRY'S MIND WITH THE NATION'S HEART"

– Reading and information –

Claude Levi-Strauss said, during an interview with the French periodical "Lire" that the Library will command primary interest for the ethnologist of the third millennium as something that retains the best of our civilisation and can say the most for the future. The French scholar had considered that a library would be the best placed for transmitting the world culture for the future. He mentioned the Library of Congress as the institution. The question posed was what feature of life might make the deepest impression on French science, thinking and culture.

It's unlikely that Levi-Strauss was wrong. At all probability it is the library which will integrate the knowledge of the present for future use. Other institutions might have a say too but the library's role, its position is optimal. Antal Szerb one of Hungary's prominent men of letters said, recalling his Paris trip between the two wars, that "It was good to be anywhere in Paris but most of all in the Bibliothèque Nationale". In a debate between Lunacharski – a leading figure of the theory and practice of socialist culture – and Chuskovski, extolling the virtues of the ideal revolutionary, the latter's hero said: "Should you have to burn Dante at the stake in order to win the victory for freedom – don't hesitate." But Lunacharski retorted: "Revolution is not an aim – it has never been. It is the means to an end. The means of creating a harmonious culture, something to nurture human strength and beauty." The 18th century encyclopaedists: Diderot, D'Alembert and their associates had already based their great work on the assumption that men are equal, so are their countries, consequently knowledge was for everyone. How? The answer was: by reading. May I add: with the help of libraries.

But isn't the traditional form of reading and the library out of date today, in the age that the Viewdata system links up the telephone with the computer for information and entertainment? Hardly so. The users of databases – which had been built up in the last decade for scientific information – were and are the scientific and special libraries. Libraries took the lion's share in the introduction and spread of bibliographical data-culture.

These databases yield information on special literature (with its varied forms, retrospective searches) while the texts of the literature itself is provided by libraries. We can, therefore, speak of primary (text) or secondary (bibliographical) information.

The Viewdata system is able to connect the primary with the secondary information for mass communication (TV programmes, video-news) and link that up with general and practical information such as weather reports and timetables. The system may be regarded as the library coming to the home.

But if it was a possible trend is it desirable? Technically speaking, yes. From the points of view of openness, intellectual stimulus and exchange of opinions – whatever matters in human relations – the trend is undesirable. In many cases, such as those who are bedridden or have no access to libraries, its "coming to the home" should be advantageous. Otherwise the system, for all its technological wonder, falls short of the infinite selection offered by libraries. Indeed it can dehumanise reading, isolate the researcher and alianate man from its natural surroundings, from the aesthetic joys of reading. In many countries today the problem is not (or not yet) related to the "library coming to the home" but the opposite: the lack of spaces in libraries.

What would the realistic cultural "model" be like that has been so much sought after and which, as Lunacharski said, has the limitless possibility of growth within a harmonious civilisation?

One of the possible answers was given by Victor Hugo who in 1850 said in debate of the legislators: "Wherever there is a piece of land tilled there should also be a book... Free and compulsory education should start with the village school and gradually... it should reach the standard of the Academie Française... It would open the gates of knowledge before everyone... and from the schools and the intellectual workshops into the libraries and to the university departments a colossal network would be formed which could radiate culture nation-wide and would give birth to new abilities and fuel devotion to vocations. In other words the State would see to it that the totality of knowledge would be available for everyone... to connect the country's mind with the nation's heart."

Reading and information to scientific work are channels of this connection. In countries not abundant in energy or in natural resources (at mid point between developing and developed) further points should also be kept in mind: 1. Of the national potential information is most readily renewable; 2. It does not lose its value but (when renewed) increases it; 3. It is a power base that can always be exploited; 4. Even when technical and material provisions are lower information for research could reach an international standard which is relatively cheap at least when compared to other costs and investments.

If the above points are promoted by the library profession of a nation then the connecting of the mind and the hearts will be a step nearer in the following years.

In: Magyar Tudomány (Central Organ of the HAS), 1981.9. 641–642.p.

ON THE EVALUATION OF INFORMATION SYSTEMS AND THE USE OF SCIENTOMETRICS

Formulation of the problem

The evaluation of the utility and the quality of information systems and processes is a widely debated problem. The concepts of efficiency, rentability, cost-effectiveness are very closely linked with the problem of evaluation. The literature of the economics of information and its bibliographies show a growing interest in the evaluation of information systems and services.

We are dealing here with the evaluation of automated information systems, as these are more numerous and more widely used both in R and D and in problem solving. The impressive number of computerized data bases is significant in itself. As Tomberg's Data Base Guide indicates the number of such data bases was 533 in 1978, that of data banks 568, and of referral services 577, amounting to a total of 1678[1].

How to choose among the large number of data bases, how to evaluate them, what kinds of criteria to use? These are difficult questions, as evaluation has to serve something, it is not a question "en soi".

But the evaluation of information systems is nothing but **an integrated part** of the evaluation of R and D. The investigations of the utility of information systems is in fact a very important **aspect** of R and D evaluation. Saying this, it is to be recognized that the evaluation of information systems has its own **specific** questions.

First of all it is the stage of the evaluation of R and D which is to be examined because of the very close links between the evaluation of R and D that of information systems.

Generally the following types of research activities are recognized in R and D[2]. **Basic research**: it is aimed at producing new scientific results, knowledge, new theories and scientific laws. Within basic research difference should be made between the so called **pure research** motivated only by the curiosity of researchers and the **oriented** basic research directed by the scientific or economic interests of a management body or of a research unit. Basic research in general is not aimed at attaining practical results. The aim of **applied research** is also to produce new scientific results but with a view to a practical achievement, with a view to applicability. **Development** utilizes the existing scientific results to produce new materials, tools, processes and systems. Development research is in general of experimental character. In development research activities the direct economic-technical results are of major importance.

Some activities are linked with R and D such as postgraduate and higher education, information processing and dissemination, collection of general data (meteorology, astronomy, public health, etc.), standardization, investigations of the finalization of the results of R and D, patents and innovations.

It seems reasonable to follow the evaluation criteria for the above-mentioned types of research activities. It would be **irrealistic** to evaluate the direct contribution of information systems to basic research. It is the **effectiveness** of such systems that can be evaluated.

We can observe also a **contradictory process**. The more theoretical and intellectual the input the less obsolescent it is in terms of research; the more its information needs are the less its measurability (efficiency, etc.). The information system as a whole can be evaluated only *a posteriori* in an indirect way. This applies to basic research. On the other hand: the less the relative importance of the information system in applied and development research is and the more the obsolescence of their information is, the more the measurability of the used information is as a part of the new products of services. This contradictory process can be formulated as follows: **more information value – less measurability**.

It means that an important part of information systems cannot be evaluated directly. Does it mean that the evaluation and the quality of a great part of information systems and services should be neglected? Certainly not. There are some elements which must be taken into consideration and thus the problem can be approached.

For the evaluation of information systems and services an **approximation** of their utility may be sufficient and **Mendeleev**'s statement, that science begins where and when something can be measured is valid only to a certain extent for information.

At present we think that the formulation of the questions of measurability and of the evaluation of information systems can be considered as a result of or an approach to the problem. And in many cases the formulation of questions can be a step forward in the process of research.

Evaluation of information systems and scientometrics

The term "scientometrics" was coined by V. V. Nalimov and T. M. Mulchenko in their monograph published in 1969[3]. Scientometrics which has got established widely since then and already has its own international journal[4] deals with the study of science by mathematical-statistical methods. The authors of scientometric papers analyse the respective interrelations of the volume of scientific periodicals, of the number of research workers, of the expenditure on researches, of the citations, and so forth. They try to study the effectiveness of groups and individuals.

We shall now specify some selected questions some authors deal with in their work[5].

We selected these questions according to the extent to which they can be related to our topic, i.e. the evaluation of information systems. The study of the topic mentioned does not take us immediately to evaluation but we think it does help us to approach it and to elaborate adequate methods. Consequently the questions to study are as follows:

– an information model of the developmental mechanism of science;
– growth of the information flow;
– statistical analysis of the content of information flow;
– citation indexes;
– statistical regularity in the system of publications;
– study of the internal connections of science by means of bibliographic citations;
– an information system based on bibliographical citations;
– scientific periodicals as communication channel;
– effectiveness of the participation of the individual countries in the world-wide information flow;
– rank of the individual countries in the world-wide flow of information as a funktion of time variable.

The views of Nalimov and Mulchenko were influenced by Derek de Solla Price, especially by his work "Little science, big science" published in 1963.

Here we have to mention that while scientometrics and bibliometrics are defined in several places as identical terms, yet there are considerable differences between them. According to a Hungarian author, bibliometrics can be defined as a method of quantitative analysis of documents used for recording, the communication of scientific knowledge[6]. As scientific publications generally serve as the media of new knowledge it is obvious that bibliometrics is an integral part of scientometrics. The subject matter of the latter was summed up by another Soviet expert, G. M. Dobrov, during his visit to Budapest in 1977 as follows:

– quantitative methods in research organization;
– quantitative methods in the economics of science;
– quantitative methods in analysing scientific information;
– applications and characteristics of science indicators. Science indicators can be considered as a new evaluation tool for science policy studies. They were established in the United States where these indicators have been published every two years since 1972.[7] Science indicators tend to turn the measurable data on research activities and achievements to unobtrusive figures, forming time-series and allowing international comparisons.

For summary

Truly objective methods which could be suitable for the measurement and evaluation of information systems, services and processes are not known, at least not by the author. In the field of science and technology information at least – and it is at the same time a crucial point – the paid consumption of information is a basic parameter. An application of methods based on new up-to-date statistics for evaluation might be a possible approach. Therefore it would be worth considering the possible application of achievements of bibliometrics, scientometrics and science indicators in measurement and evaluation within the qualification sphere of information systems. This perhaps does not belong to the less promising methods. It is the FID/RI that could initiate such international researches.

NOTES

1. A. Tomberg: EUSIDIC Data Base Guide. Oxford, 1978, Learned Information, Inc.

2. Ch. Freeman: Measurement of R and D and other scientific and technical services in the economics of industrial innovation. Harmondsworth, 1974, Penguin. pp. 311–387.

3. Naukometriya. Moskva. Izd. Nauka. 1969.

4. Scientometrics. An international journal for all quantitative aspects of the science of science and science policy. 1978. – Elsevier Scientific Publ. Co., Amsterdam – Akad. Kiadó, Budapest. Edited in the Library of the Hungarian Academy of Sciences.

5. V. V. Nalimov, I. M. Mulchenko – in Naukometriya, Moskva, 1969.

6. Pál Györe's study in Tudományos és Műszaki Tájékoztatás, 1974, 8–9, p. 598.

7. See Science Indicators 1972, 1974, 1976, National Science Foundation, Washington, 1973, 1975, 1977. Also the serial publication of the Library of the Hungarian Academy of Sciences "Informatics, and Science Analysis" is to be mentioned here. This serial is produced by a specialized Research Group of the Library.

In: Criteria of the quality of information systems and processes. Study Committee FID/RI. FID Publ. 591. Moscow, 1981. 77–81.p.

PROBLEMS AND PERSPECTIVES OF NETWORKING IN SOCIAL SCIENCE INFORMATION AND DOCUMENTATION IN HUNGARY

1. Raising the problem

The problems of social science information and documentation (henceforward SSID) as well as some questions of its organization into networks are by and large identical with the body of knowledge and experiences generally known from the related literature. Obviously, there are also specific problems in the same way as there appear some features peculiar to every country, coming from its particular cultural standard, its traditions and the level of economic development. Therefore, the general approach to the problems of SSID will be omitted here, instead I will endeavour to point to what is peculiar to SSID in Hungary, emphasizing the practical experiences. Furthermore, despite the fact that the author of this paper may well represent the Hungarian SSID effort in more than one capacity, this is not meant to be an official statement; it is to express his personal opinion.

2. Antecedents

It is not an exaggeration that the development of SSID in Hungary and its organization into network have a past of at least two decades, at least. In reality efforts to organize SSID can be traced back to the early fifties. Without going into the historical details of SSID in Hungary, it seems necessary to point to some phases of its development.

In one form or another, the large social science libraries in Hungary have always carried on such activities – at least on an embryonic level – that are contemporarily known as SSID. Its forms were highly varied, and fortunately enough have remained so; the products of information were only partly regarded as documentation, in fact they were mostly routine library information services and publications. Lists of new acquisitions, current awareness cards in specific fields, thematic (special) bibliographies, selected titles of articles and similar means of information have made orientation in social science literature possible. Then, in the very beginning of the fifties, the Hungarian Academy of Sciences initiated the establishment of special documentation organizations – and this has already been a network – of which social sciences were not left out either. The main emphasis even then was laid upon the natural sciences and technology. A central institution, the National Documentation Centre and two specialized documentation centres (for the social sciences) were set up: namely, the Economic Documentation Centre and the Documentation Centre for Literary Studies. Of the latter two, the Economic Documentation Centre lived longer to merge later, during the second half of the fifties, with the Library of the University of Economics. A minor part of its functions was taken over by the then newly founded Documentation Services of the Institute of Economics of the Hungarian Academy of Sciences.

The above mentioned institutions were engaged in several activities which were not covered by libraries at the time. Thus, for instance, they were dealing with such works as the Hungarian translation of UDC and its introduction to a number of special institutions, the registration of translations and translation from foreign special literature, the preparation of digests and analytical reviews. In the Cold War years, all this had its own particular significance, but these early documentation works did not only give the specialists an insight into Western scientific literature, but they also had their uses from quite another aspect: they provided to a considerable extent, direct access to Soviet literature, necessitated by a lack of knowledge of the Russian language.

Set up early and ceased too early, this documentation organization started, even if in an immature form, what is now called networking. This networking was not based on a division of labour; it was rather an initiative to help the services of special documentation centres reach the users through other institutions as well. Therefore, relations were established and a sort of methodological guidance and exchange of experiences were developed among the institutions concerned, and they mutually took over and exchanged their respective documentation services.

3. Beginning of the SSID networking

After World War II two principles became predominant right at the beginning in the organization of the slowly recovering Hungarian library life. One, was the unity of library affairs, meaning that both special and public libraries were understood to be part of a uniform national library system or network. The other was the network principle itself, which groups the libraries of various types or working under the supervision of various national bodies either from the aspect of administration (maintenance) or from that of subject fields (e.g. economics, law, etc.). Thus the starting points for networking in SSID were given even from a legal viewpoint. However, as shown by the Hungarian experiences, it is easier to regulate something legally – networking in this case – than to give it meaningful content and run it efficiently.

At the end of the fifties, a collaboration among three institutions resulted in a Hungarian documentary publication in the social sciences which was the first to be based on real cooperation and which, in many respects, already bore the signs of up-to-date documentation. The title of this publication was "Közgazdasági és Statisztikai Irodalmi Tájékoztató" (Information on Economic and Statistical Literature), edited jointly by the University of Economics, Central Statistical Office and the Library of the Institute of Economics of the Hungarian Academy of Sciences and published by a professional publisher, the Publishing House for Economic and Legal Literature. This periodical was in fact stopped because of its having been in deficit.

During the sixties, SSID activities were increasingly intensified in several big social science libraries and also in the information services of the Academy's research institutes. The political consolidation in Hungary in the sixties and the related economic upswing induced an increase of interest in social science literature. The international situation had also a favourable effect in the same direction. Various cooperative efforts were made to satisfy the growing interest mentioned, but it appeared before long that the greater the possibilities of SSID to satisfy this interest, the more increasing were the autharchic tendencies within the institutions. In other words, the more favourable possibilities were not followed by their equally favourable exploitation. The sometimes peaceful, sometimes discordant coexistence of cooperative and autharchic tendencies have benn characteristic of SSID networking in Hungary up to now.

4. Causes affecting adversely SSID networking

It was in the early seventies when the cultural government authorities – which have been acting up to date as the nation-wide professional supervising body of library affairs – found that time had come to develop and organize SSID into a network on the principle of centralization. To achieve this, a secretariat, made up of a few experts, was set up within the Centre of Library Science and Methodology. This secretariat partly organized surveys, partly made efforts to bring together the libraries concerned. This activity could be considered ground-work if only because during this work the introduction of up-to-date information technology the use of computers for SSID was already proposed. In addition, the rather active Hungarian participation in international SSID efforts had made its effect felt in the related Hungarian works from the beginning. Among these are several years' work in ICSSD (within the so-called Meyriat-Committee), and activities in FID/C3, the shorter or longer work of some Hungarian experts within international professional organizations, and so forth. All this resulted in the tendency to increase interest in SSID – which had commenced as early as the sixties – taking a more palpable shape. It should also be noted here that the new economic mechanism (reform of economic control) – which is still going on in these days – has increased the "hunger for information" in general, and within this the requirements were set for SSID.

At the same time there came about a singular situation: simultaneously with centralization-minded conceptions and cooperations developed, there were also decentralization-oriented conceptions, following from the economic reform, which were contrary to cooperation and coordination. According to this, the institutions were supposed to provide for their own information to which they were supplied with appropriate financial means, (thus they were inclined to cooperation – i.e. networking –) to the extent as they felt it to be in their own interest. So, why would any institution, having all financial, organizational, personnel and other possibilities for complete independence, feel it necessary to cooperate with others? Why would it give up any of its total or virtual independence if not necessary? This question may well be put both in the case of people and of institutions. Indeed, certain coordination secretariats could not proceed further than the initial groundwork owing to the above-mentioned causes and last but not least to the lack of financial means.

Networking everywhere in the world – in Hungary as well – implies one or other form of coordination and cooperation. Someone has to organize the cooperation, eventually a division of labour and to provide for its regulation. Among the factors contrary to coordination and cooperation is the reason that usually the real backer of coordination is the coordinator himself, while the others – who are "being coordinated" – are less enthusiastic about other institutions interfering with their affairs. I think this reason accounts for the negative tendencies in networking.

5. Attempt at an organized cooperation (Working Group for Social Science Information)

What has been outlined above has, I think, a general validity which includes Hungary. The recognition which led the concerned professional authorities and bodies, as well as quite a few representatives of the Hungarian SSID further, can be summarized thus: SSID networking should be based on mutual interest. Such motivation must be found and made loose enough not to make the participants feel to be guided administratively, at the same time they should positively benefit from their participation in networking. In other words, all the presumed effective guidance, good will and enthusiasm are insufficient by themselves for networking.

In the late seventies, the supreme science policy body of the government, the Committee on Science Policy (being presided over by a deputy prime-minister) with the effective help of the Hungarian Academy of Sciences, set up the Social Science Coordination Committee (SSCC for short), to plan and harmonize social sciences research works going on in Hungary. This committee – jointly with the Ministry of Culture also responsible for the professional supervision of libraries – began to oversee the development and modernization of SSID. The Working Group for Social Science Information (WGSSI) was set up to act as its advisory body. Today this Working Group also helps the competent bodies of the Ministry of Culture. As it has worked out over the years, SSCC exercises direct professional supervision over WGSSI and has been providing – practically up to now – for its finances (more details later). The president of SSCC has invited the heads of large social science libraries to participate in the work. The WGSSI has 15 to 20 members. This is the adequate size of membership still capable of working flexibly and since it includes the leading staff members of basic SSID institutions too, it can provide for an appropriate institutional background necessary for all kinds of organizational work. To put it another way: we endeavoured to combine professional knowledge with the representation of corporate interests in one and the same body. If either of them is missing in the guiding-organizing activities of networking, efficient work seems impossible. Afterwards there has been a period of many experiments and halts which in fact continues even today although some achievements are already appearing.

WGSSI had from the outset sought to maintain informal connections as well and to hold sessions as rarely as possible, offering, however, many opportunities for discussions. This method of working did not prove unsuccessful. WGSSI had interdisciplinary analyses prepared on the state of SSID in individual branches of the social sciences (sociology, economics, literary studies, etc.) and these analyses were put on the agenda of certain sessions involving external experts where they were discussed and then submitted to the competent scientific committee of the Academy asking for its official standpoint on the given matter. It was the assumption – and part of it is still valid – that the coordination of SSID should not come from one particular centre, but each major discipline should have its own base institution within the SSID network. Thus, for instance, in the field of legal and political sciences the Library of Parliament seemed to be the most suitable institution, likewise the Municipal Ervin Szabó Library in sociology, or the National Pedagogical Library in the broad field of education and so forth. The main criterion for designating a base institution were: the professional (scientific) significance of the given collection and the quality and possibilities of its services. The "quasi coordination" of the branch-specific SSID base institutions is carried out by the Library of the Hungarian Academy of Sciences, partly as an interdisciplinary institution, partly as a significant document- and servicing-basis taking into account the special libraries of the Academy's research institutes, too (libraries of research institutes for history, sociology, economics etc.). The undertaking of this task on the part of the Academy's Library was and still is justified by the fact that it is representing the Academy and Hungary in two international social science programs. The first program is MISON (International Social Science Information System of the Socialist Academies) and the other is ECSSID (European Cooperation in Social Science Information and Documentation, an East-West program). The bearers of the burdens of development and modernization are the branch-specific SSID base-institutions in Hungary while the Library of the Hungarian Academy of Sciences is responsible for the organizational and financial support of these efforts.

The branch-specific SSID analyses and generally the conception of WGSSI assumed that the information base-institutes are oriented from the aspect of science policy by the professionally competent research institute or scientific committee of the Academy. This assumption seemed to be feasible as it was to connect research with SSID, but in practice it turned out to be a failure. It could also be said that the bodies or institutes supposed to give the disciplinary scientific orientation were not adequately motivated to fulfil this task. But it also became clear that SSID networking cannot be brought about by mere conceptions, however impressive and convincing they may be. Thus WGSSID finally came to the conclusion that an interest-oriented organizing force should be found around which SSID can realistically be organized and filled with real and meaningful content. The introduction of up-to-date information technology in SSID, along with a cooperation based on computerized information processing seemed to be such a realistic content element.

6. Computer-based SSID

Owing to the economic circumstances and the level of technological development in Hungary, it seems to be rather feasible to create machine-based literature processing and data bases in the form of cooperation. Machine-based data processing which started relatively late, generally meant a lack of experts as well. This particularly applied to the processing of printed information (literature), and within this mainly to SSID. Therefore, the initiative of WGSSI – namely that after several experiments, plans, conceptions, SSID and the organization of related networking should all be based on computerized literature processing – was received favourably by the institutions concerned. To achieve this end, (with respect to the scattered and uneven state of financial and technological means and experiences) the tasks to be performed had to be summed up in a sound conception which could take into consideration the real needs of scientific and economic life. With this, WGSSI wanted to avoid, as much as possible, the appearance of incidental and random developmental tendencies. Therefore, utilizing the experiences gained by various libraries so far (e.g. Library of the Parliament, Library of the Central Statistical Office and so on), and analyzing both the real demands on machine-produced information and its possibilities, WGSSI prepared a skeleton-program of SSID development for SSCC. After several discussions this higher science policy body accepted the WGSSI's skeleton-project, along with its financial consequences. It was obvious for all that any conception without adequate financial means is nothing but an illusory objective. The government's Committee on Science Policy allotted an amount to WGSSI, which sum, compared to the country's financial position was not insignificant to perform the given tasks. The Library of the Hungarian Academy of Sciences was entrusted with the handling of it. The accepted conception established that the computer-based SSID should be started in a few priorized fields. These fields were marked out by WGSSI on the basis of scientific and economic considerations and of the SSID work done so far. They are: economics (including statistics), sociology, legal and administrative science, and, at a later date, pedagogy. (Here I note that in the field of science policy and research organization some computer-assisted scientometric researches are going on within the Library of the Hungarian Academy of Sciences.) About half of the allotted sum was spent by WGSSI on data recording machines to be run by the base-institutions of the three fields mentioned. From the remainder, the costs of intellectual works necessary to computerized SSID and to its input were covered by WGSSI. Thus, among others, branch-specific and network systems planning costs, costs of document analyses, bridging linguistic gaps (translations) and the like are covered. The work described above is still going on.

It has proved true that with the institutional-disciplinary interests given concrete form, the interest in cooperation and networking has increased. Instead of declaration and incitements institutions were given financial and related professional support and these two have had a favourable joint effect on their "frame of mind". Computerized SSID in Hungary is only taking its first steps in most fields, nevertheless it has embarked on a road that proves to be practicable. This is not to mean that all difficulties in SSID networking have been eliminated since there are and will always be hotly debated questions, not infrequently of prestige nature, but the basic consent and the willingness are given. In one or two fields the branch-specific SSID networking shows somewhat more marked outlines as in the fields of legal and political sciences, economic policy or statistics. Not insignificant preparatory works have been done in sociology. Initial steps have been taken in the field of pedagogy. In some of the other fields demands have come from the research community (history is a case in point).

It can be expected that organized into network, computerized SSID in Hungary will go on along this line, and its automated services – combined with the related international services – will be able to provide more and more help to research and practice.

* * *

This is, then, the picture – or rather its sketch – that seems realistic of the state of SSID and its networking in Hungary.

In: FID studies in social science information and documentation. Networks and networking in social science information. FID Publ. 606. Budapest, 1984. 77–86.p.

INFORMATION COMMUNICATION SYSTEMS AND THEIR IMPACT ON SOCIETY

Communication in society has gone through thorough, even revolutionary changes in the course of history.

Three big phases can be distinguished, oral, written, and printed communication. The first is a part of auditive communication, the two last ones of visual communication. The appearance of the wireless in the first half of our century proved the social importance of auditive communication, in spite of the predominance of visual mass communication acquired first by the silent movies, and later by the sound- and colour films. I am referring here to all information communication as a macro-social phenomenon including organized information, if it is not person to person.

Today, after the victorious march of television (the most developed audio-visual medium) the triumphant conquest of informatics, we see the rise of a fourth great phase of evolutionary change in information communication, i.e. telematics, or the viewdata system. Viewdata as a collective notion includes television, the telephone and the computer.

This is not the proper place to describe the technical details of the viewdata systems. Incidentally, the technical side is known well enough. But the social side of the phenomenon should interest us more. In other words, what can be expected of viewdata in relation to the already existing cultural and scientific media? To what extent will this information communication category change the use of the traditional channels of communication: printed media (books, newspapers, the press) and audiovisual media (television, film, radio)? What can the influence of viewdata be on scientific research, on scientific information services, on the methods and techniques of becoming educated, gathering information, collection of knowledge, or the spending of leisure, the time spent at home?

I shall try to sketch hereunder only in broad outlines the characteristic features of the two principal categories of the transmission of information and knowledge in the future: viewdata, as a collective concept for the techniques of communication points, and the library, as the collective notion for traditional reading.

In my opinion, we shall witness a process of integration of information and a regrouping of knowledge which will rely on three pillars: electronic communication (posts, telephone and telegramme; satellites, etc.), the computer, and the library. Viewdata should become the catalyst of this process.

Let us then see what the viewdata system can offer to libraries. I shall already here indicate my conclusion: services of the two types of communication will be complementary and not competing against each other: there will be a reciprocal stimulation process without negation or exclusion.

Here are my arguments in support of this preliminary finding.

Library services	**The "viewdata" services**
1) SOURCE (complete) The source of knowledge has no (chronological, historical, linguistic, thematic) limits.	1) SOURCE (incomplete) The source is restricted, very limited as far as integral texts are concerned, and nil in respect of manuscripts, old books, rarities, etc.
2) DIFFERENTIATION OF SERVICES (complete) The services can be differentiated, without any limitations, according to social standing, age, occupation, knowledge of languages, etc.	2) DIFFERENTIATION OF SERVICES (incomplete) It is possible to differentiate services only relying on limited sources.
3) RECIPROCAL USE (yes) The library can be used unlimited.	3) RECIPROCAL USE (no) Viewdata can be used only with great limitations (of sources).
4) RETROSPECTIVE RESEARCH (yes) Unlimited.	4) RETROSPECTIVE RESEARCH (no) With limitations set by the input of sources.
5) COST OF SERVICES (inexpensive) These are public services; the majority of services offered are free of charge.	5) THE COST OF SERVICES (more expensive) It appears that the cost of these services will be higher than that of libraries, in spite of the fall in prices due to "public relations" and to technical costs.
6) SERVICES AND THE WAY OF LIVING (harmonious) The use of the services can be included harmoniously in anybody's way of life.	6) SERVICES AND THE WAY OF LIVING (harmonious) Harmonious inclusion is not without restrictions: those of the price of services.

7) INDIVIDUAL AND COLLECTIVE USE
(collective)

Collective use offered by libraries with a stimulating atmosphere, possibilities of browsing though new acquisitions, consultation of catalogues, is irreplaceable.

8) SOCIOLOGICAL EFFECT
(human relations)

The sociological effect is clearly positive, as human and professional relations and contacts are created among individuals.

9) DISPLACING THE READER'S ACTIVITY
(active)

The use demands a certain activity, sometimes uncomfortable (moving to a distance, noise in the library).

10) AESTHETICISM
(yes)

Handling of books, and of documents in general in the libraries makes the manifestation of a certain aesthetic effect possible (the typography, the binding of works, etc.)

11) ACCESSIBILITY
(unlimited)

Access is unlimited, free of charge (or nearly so), democratic, and has a socializing tendency.

7) INDIVIDUAL AND COLLECTIVE USE
(individual)

The data bases penetrate "privacy", the homes of the users; the latter are deprived of a stimulating atmosphere.

8) SOCIOLOGICAL EFFECT
(alienation)

This has a contrary effect: the use of viewdata dissociates the individuals from each other, and pushes them towards solitude.

9) DISPLACING THE READER'S ACTIVITY
(passive)

The use has a certain effect towards passivity, which however can be very comfortable as it introduces the library, at least in part, to the home.

10) AESTHETICISM
(no)

The print-outs of the databanks make writing uniform, and they have no aesthetic effect.

11) ACCESSIBILITY
(limited)

The access is limited by the higher cost, and there is a tendency towards isolation in this domain.

Leonardo da Vinci conceived that in some respects the ideal proportion of man would be to have four arms and six legs. Viewdata can provide some elements of this ideal proportion. Culture and science are imperfect by definition, but are always evolving. Consciousness of this evolution implies the recognition of the efforts specialists in informatics, librarians, professionals of the book trade, etc., i.e. generally speaking of educated people.

In: Schéma et Schématisation, 1983. No. 18. 52–54.p.

RESEARCH AND SSID SERVICES, TENDENCIES AND CHARACTERISTICS

IV. ECSSID Conf. Athens, 21–24. October 1984.

Co-author: Jiří Zahradil

1. A few words about the classification of and in the social sciences

The concept of the social sciences is interpreted in different ways, which makes it necessary to give at least the broad outlines of what we mean by the social sciences. There is no doubt, of course, that they deal with the development and history of human society and man's place in it. Nevertheless, it is by no means an easy task to name each particular branch belonging to the social sciences. To illustrate the difficulties of classification – and as an interesting case – let's take the example of history. In his time Marx believed that history was "the only real science". In the sixties, however, at a UNESCO special experts' conference, some of the excellent social scientists who took part in the conference (Lazarsfeld, Piaget etc.), had a heated debate on what history really meant. One of the opinions was that considering history we cannot in fact talk about a branch of science, but a dimension (one of the preparatory conferences of the so-called Auger-report on the social sciences).

According to western interpretation, the branches that belong to the social sciences are only those, that study society specifically, such as economics, sociology, political science, law and public administration, and so on, whereas philosophy, history, psychology and other branches of science that form human consciousness, belong to the humanities.

In the socialist countries by social sciences we mean a complexity of knowledge that comprises humanities as well. So the nomenclature of the social sciences in these countries includes – without the list being complete – the theoretical aspects of socialism, philosophy, economics, sociology, statistics, demography, history (including archaeology, the history of art and ethnography), political science, linguistics, the study of literature, pedagogy and the theoretical problems of culture.

Concerning the problems of classification, we should mention the revision of social science classification that has been carried out in east-west UDC cooperation under FID C/3 for 25 years and the thesaurus studies made by ECSSID, in order to throw light on the problems of "what belongs where" and "how to interpret things" from the point of view of SSID.

UNESCO, both in its activity and structure, differentiates between the social sciences and the humanities, and so does the International Bibliography of Social Sciences, the most important international bibliographic venture sponsored by UNESCO.

These two different interpretations of the social sciences – the integrated socialist version and the western classification (social science and humanities) – and their impact on SSID, do not impede international scientific cooperation in the SSID. A good example of this is ECSSID.

2. The three main functions of the social sciences

A closer study of the functions of social sciences will show – regardless of the above mentioned differing interpretations – that the social sciences as a whole and each of their different branches have three main functions. The first one is ideological, the second one is the exploration of social reality and the third one is the preparation of decision making. The social sciences do not exist as "pure sciences" (l'art pour l'art, art for art's sake). Different ideas are always, in some way, reflected in social life and the social sciences are influenced by scientific research. The three above mentioned functions of social sciences rarely appear independently from one another, they are usually interrelated. However, as we shall see later on, the differences between the three functions are manifest in the field of information, namely in the different demands of those who use information.

Whether we regard the social sciences as a whole or divide them into the social sciences and humanities, each branch is characterized by its own, particular way of work, which – as we shall explain later on – exerts strong influence on the process of information and not only among the particular branches of the social sciences but among the natural and the technical sciences as well.

In this respect let us refer to the application of physics in archaeology, the growing use of cybernetics in social sciences, for instance in lexicography, to say nothing of economics, or sociology that have, by now, their "traditional" fields of application. Today, there is practically no field in the social sciences, not even museology or archival research, that is left out of this process.

The three functions of the social sciences – along with the recent improvement of measuring methods and the new technology in information – lead us to draw certain conclusions concerning the future development of SSID. This, however, cannot be done without examining the different, previous periods in the development of SSID.

3. The three historical periods of scientific information

The formation and application of scientific information is governed by the same rules in the social sciences as in other fields of science and in society in general.*

In his theories on the surplus value, Marx explains that the competence of the existing population is an invariable precondition of all production i.e. the principal

---

* For further information on this chapter see György Rózsa "Scientific information and society" (The Hague, Mouton, 1973)

accumulation in economy: this is the most important result of previous work but it exists in current work itself. This "principal accumulation", that of production experiences, – knowledge systematized into science in the 18th century – is being summed up and forwarded through and by the "collective memory" of mankind – scientific literature.

This principal social function of scientific literature reflects the process of science becoming a productive force and is also one of the most essential parts of this process. Both sides of this process assume their final shape in the scientific and technical revolution unfolding in our age and this in turn by that of the sciences. Hence it follows that in the last analysis the material production of society is the technological application of sciences.

The role of knowledge ("competence" of society) and its application in the development of the forces in production have evolved in three major historical stages, within which the transformation of scientific writings into specialized scientific literature takes place along with the increase of scientific literature, the appearance of scientific information and their concomitant problems.

In the first period, that is, up to the industrial revolution, we cannot talk about the social utilization of science as it is understood today. The organized conscious and wide--scale application of science were not yet possible owing to the relatively underdeveloped state of the productive forces. Science was studied individually, on a small scale, and the chief collective form of communications was the book. Verbal communication and correspondence (invisible college) were also of significance, but that of periodical publications was scarce. Scientists working without reference to one another were able to master the existing knowledge in their field, but this "collective memory" was not disseminated beyond the walls of libraries.

The second period, lasting from the Industrial Revolution to the beginning of the scientific and technical revolution unfolding today, is, in fact, the first stage of an organized and conscious application on a social scale of science as a productive force. Sciences become specialised, new disciplines are born. Owing to the tremendous development of the forces of production, the isolated, individual scientific work is being replaced by scientific work on a large and wide scale. As formulated by the Communist Manifesto: The intellectual creations of individual nations become common property. National one-sidedness and narrow-sidedness become more and more impossible, and from the numerous local literatures, there arises a world literature. Beside the book, the journal gradually becomes the primary source of scientific communication. A direct survey of existing knowledge in a discipline by any single individual becomes impossible, a mediator therefore develops between scientific "world literature" and research – it is first the bibliography, then the documentation.

The third – that is the present – period is characterized by the scientific and technical revolution. Both science and its assessment become direct productive forces. According to estimates, the number of scientists in this last period amounts to 90 % of the total number of all the scientists having lived up till now. It follows that the organisation of work should be more in compliance with this number. Big research institutes are created that work on complex tasks, team work takes shape. The most important fields of research assume a trans-disciplinary character. These phenomena are wholly manifest in the social sciences even though there is some delay in time when compared to the natural and the technical sciences. The tremendous increase in the number of scientists is reflected in the increased number of publications, the final products of research. In the social sciences the final product of research is practically always a publication. Data on the increasing number of publications have been published by several institutions. According to the UNESCO/Auger-report, the number of periodicals has increased as follows:

Early 19th century	100
1850	1,000
1900	10,000
1960	100,000

The 1987 edition of ULRICH, the world list of periodicals takes account of 96,000 titles, but the list is not all embracing. The Soviet scientist Dubinin estimated in the 1960s that the number of scientific publications had doubled in every 15 years. Although this process has slowed down since one is reminded of another estimate, by the American professor Kent, whereby a days output of print: books, papers, reports, documents requires a thousand million pages every day. This estimate was modest even in terms of the seventies.

However, certain effort is being made by the authors of publications, that is the researchers themselves, to control this "publication boom". There are discussions on what to publish and how to publish.

The efforts to enforce the idea of "publish or perish" are also well-known. At the same time there is a tendency to limit the flow of publications. It was Bertrand Russell's idea – as long ago as 1948 – to have periodicals that publish abstracts only and to keep the full manuscripts in regional centres – but this remained an idea. So today – and probably for some time on – we have to accept scientific information as it is, with its ever improving technical possibilities.

4. The three main source-orientations of research in the social sciences and the SSID particularities

The interest in publishing – along with some other factors – has made some researchers overestimate the importance of certain publications. In the course of a survey conducted by the Czechoslovak Academy of Sciences among 1000 researchers working in the social sciences, the different sources of information had to be classified in their order of importance. The results were as follows:

1) periodicals	– 47 %	of people involved in the survey
2) books	– 30 %	– „ –
3) reports of research, grey literature	– 13 %	– „ –
4) other sources	– 9 %	– „ –

These are average figures for the social sciences as a whole, but there are no large variations in this order between the different branches of the social sciences: 40–55 % for periodicals, 17–38 % for books, 9–23 % for reports and grey literature. The order of importance itself was the same in all the branches. Another result of the survey shows the obsolescence of the value of information as a function of time (the half-time of information obsolescence), which means that a given piece of information loses 50 % of its value. Investigations of this kind are common in the natural and the technical sciences, but have been less frequent in the social sciences. (See the investigations by Maurice Line or the survey at University of Bath.)

The Czechoslovak investigations have also proved the assumption that the obsolescence of information in the social sciences is slower than in the natural and in the technical sciences, but much faster than it is generally believed. It was found that the half-time of information obsolescence in the social sciences of the Czechoslovak Academy of Sciences was 3.92 years. Only 25 % of the information needed in the social sciences was "over ten years old".

The particular needs of the different branches of the social sciences no doubt influence the half-time of information obsolescence in a given field. The branches influenced by the scientific and technical revolution and the accelerating social mobility, especially economics and sociology, rely to a larger extent on "fresh" information than, for example, history.

The large variations in the different half-periods of information obsolescence between economics (2.28 years) and history (6.49 years), between philosophy and sociology (3.7 years) and literature (8.48 years) show the considerable variations that exist between the different lengths of time each source of information can be used. This calls for the specific treatment and handling of the different sources of information as far as collection, storing and selection are concerned. The concept of "social science information" is a kind of necessary simplification rather than the exact description of the subject which is far more complex both in its contents and function.

When evaluating the above results of half-times, the limited sampling possibilities should also be borne in mind.

The different branches of the social sciences may be divided into three basic groups and one mixed group, according to the different written sources their research activities are based on. In this way we distinguish:

a) special literature-oriented branches
b) factographic-oriented branches
c) data-oriented branches

and a group comprising the three types of orientation, mixture of orientation a–c branches.

Written and printed sources are of basic importance in social science research. In certain branches daily newspapers and in others literature may be just as important as scientific or special literature. The study of printed texts is not always the basis or starting point of research. There are other, primary sources of research that have become written sources only after they had been described in writing. Such sources are, for example, manuscripts and archival materials in the study of history or literature. Other primary, non-written sources can be found in archaeology too, for example buildings, various objects of historical interest or the remains of skeletons. Of course most of these sources themselves have been discovered with the help of previously made descriptions, maps or other written material. As further examples of sources of information we should mention various objects of art or objects of artistic value such as the cave drawings of Altamira or the scrolls of the Dead Sea, that had to be deciphered by scientists. Sound recordings and folk song recordings are particularly important documents in ethnography, whereas recorded speeches and audiovisual material in general are valuable documents in the study of history.

Several branches of the social sciences rely, for their research, on data-oriented information such as statistical books, surveys, data archives and so on.

The source-orientation of different branches of the social sciences is shown below:

Branch of science	Orientation (combinations)
history in general	a, b, c
history of art	b, a
archaeology	b, a, c
economics in general	a, b, c
political economy	a, c
business economics	c, a
banking	c, a
sociology	a, c
philosophy	a, c, b
law	a, c, b
literature	a
linguistics	c, c, b

In practice the different source-orientations are never manifest in a chemically pure form. The source of research is sometimes one source and sometimes another one, or different sources at the same time.

We have so far pointed out the three main functions of the social sciences (ideological function, exploration of social reality, preparation for decision making). We have also discussed the three main source-orientations in social science research (special literature-orientation, factographic-orientation and data-orientation). Considering these two basic factors from the point of view of SSID, we may trace the process of the diffusion of scientific information in cognition.

5. The three phases in the cycle of cognition in research and the SSID

Without trying to suggest that the number three has some kind of cabbalistic force, it seems that after the three functions, the three historical periods and the three main source-orientations of the social sciences, we shall now study the three phases in the cycle of cognition in research from the point of view of SSID.

We believe that the cycle of cognition can be divided into three phases: discovery, evaluation and description and finally, publication.

The discovery element may be the result of special literaure-oriented, factographic or data-oriented sources, but in most cases it derives from a mixture of all these sources. The evaluation and description are related to the three functions of the social sciences (ideology, exploration of social reality and preparation of decision making), and the publication is the summing up and the synthesis of the entire work and it presents the results achieved "for approval" to scientific and public opinion. Without their criticism we cannot talk about scientific output or results as there is simply no way of measuring this type of activity.

The graphical illustration of the whole cycle would give a spiral composed of constantly reoccurring elements. Process of creation: discovery → evaluation and description → publication → new discovery → new evaluation and description → new publication... and so on, endlessly. It is the publication element – when its results are such – that finally has direct influence on science and decision making. The upper part of the spiral often represents the extended sphere of knowledge of one and the same phenomenon. The spiral is all the more a good form of illustration because it is an unfinished ouvre that refers to the infinite character of human knowledge. Social practice and science together prove the correctness of the new findings.

6. Seven supplementary remarks and a "bon mot" in conclusion

The "information service" provided by libraries in the social sciences historically preceded those provided by the technical sciences. The importance that people in ancient times attached to libraries and the information stored by them, is easily seen in the light of the flames of the Alexandrian library. But several centuries later it was in another library, in the British Museum, that one particular load of information of our age – even if it is not the most effective one – The Capital was written. Which highly up-to-date scientific information system operated by telecommunications technology or automated technical literature system has ever caused a similar "information boom" in the world? We do not, however, wish to make scholarly comparisons nor do we strive to reveal the mysteries of information in this paper. There are thousands of publications on this topic as it is. We should merely like to stress what history seems to prove: the notions "hard" and "soft" should be dealt with carefully even in the field of information.

Our main intention in this paper was to discuss certain characteristics and aspects and the historical development of social science information, but instead of opposing these to science and technology information, we wanted to point out the differences within the whole problem of scientific information, in the spirit of "diversity in unity". An excellent study has been published recently in the Journal of Documentation (1984. no. 2. pp. 94–119.) – a study based on a bibliography of 112 items –

on the present state and prospects of social science information. ECSSID conferences (I–IV., but in particular I. and IV.) help to review the European and general international development, but in some cases we get quite detailed information on the situation in certain countries. We should not forget Madame Hogeweg de Haart's review on "the state of the arts" either, which was published in 1980 (Characteristics of social science information. FID/SD, Discussion paper, 1980/2.) and which gives an excellent review of this subject, at least as far as the western approach is concerned. (Among its 64 references there is only one from a socialist country.) No doubt a similar report will once be published on social science information in the "other" (the socialist) part of the ECSSID.

Owing to the existence of these excellent reports, we think it inopportune to discuss questions such as SSID-technology or usage-analysis, on which we have plenty of information at our disposal. Only in the Outlook-Annex of this paper shall we mention a few points that, although not new, represent our own ideas in this field.

However, hic et nunc (here and now) we should like to discuss certain practical aspects that complete the theoretical ideas we outlined in previous chapters. These are as follows:

- The SSID will only be able to perform its tasks fully, if it can ensure the complete (primary) text, that is if it is based on library services (as well).
- For certain fields of social science research – humanities in particular – the main source of information will always be the library, (the study of old books and manuscripts, etc.). Nevertheless, computer technology is an important, supplementary aid in this field, too.
- Direct information is an integral part of social science research, so SSID cannot be alienated from research.
- From what we have said so far it follows, that the widely spread computer euphoria does not mean a reduced demand for reading in the social sciences; it is reading that we find at the top of the spiral of knowledge, too.
- The users of SSID are not only the researchers, a "circle of privileged people", but the participants in social and economic fields and – with the intervention of mass media – the general public as well.
- This potentially wide-range usability of SSID suggest that it would be expedient to make the users cover directly – to some extent at least – the cost of investment. This would also enable the assessment of actual demand for SSID.
- In the process of discovery-cognition, the factographic and data sources become part of the results of scientific research, they are compiled and diffused as new knowledge.

And finally let us conclude these theoretical (hopefully not theorising) considerations – by way of excuse – with an anecdote, which may not really be suited to a serious topic like this. There are two speakers at a meeting. The first man rattles off his speech from a paper, the other talks without notes and makes a spirited speech. Someone in the audience says to his neighbour: "What a speaker! He pretends to be an orator yet he is illiterate, he can't even read!"

The authors of this paper hope that their readers will not attribute their somewhat unusual approach and freedom of style to their complete ignorance.

Annex: Outlook

Development in the world today and in the SSID as well points towards the creation of large, international data bases specialising in different branches of science, from which information can be obtained on line by the main users. International cooperation is based on a unified data processing technology and moreover on a compatible form of data recording structure and on a unified system of information retrieval (a simplified, on-line retrieval language). The further development of electronics and telecommunications is decisive in this field. We may also reckon with the further growth of the capacity of memory units and the decrease of data searching time. With the extension of computer networks an increased number of data bases will be linked to one another. All over the world small computers are linked to large computers which operate the data bases by large capacity communication channels (relay transmitters, optical cables or satellites).

There are significant changes concerning access to complete texts of information sources, and a very fast development is to be expected, e.g. CD/Rom-system, the Telefax-system etc.

The new technology of SSID means that a new sphere of knowledge must be studied and acquired by both SSID staff and the users of SSID (education and training). New structures of work are created in information. Complex, inter-linked units are born which enable the computerized information processing in international systems and the availability of the complete texts of necessary documents.

One of the authors of this paper (Gy. R.) believes that the rate of this development depends largely on the rate at which computerized information in SSID becomes a marketable commodity which has to be paid for (economic and sociologic information for example).

Socialist countries today make an effort to develop SSID within the sphere of social services, as it is the case in the MISON-system. Obviously this will be necessary for some time.

It seems that with computers the main problems in east-west SSID cooperation are not so much of a technical character, but are caused by the differences between the natural retrieval languages (English and Russian). There are also problems with differing notions, lexica and thesaurii.

In cooperation with other international organisations financed by UNESCO, with the FID, the CIDSS (the so-called Meyriat-Committee) and along with the international bibliographic research, the ECSSID is a suitable organisation for promoting international cooperation, which is the main trend of development today. If it does so, the ECSSID will not only fulfil its tasks, but it will become part of a great social achievement as well.

That is the utmost we can expect of an organisation.

In: Place of Information in the Global Problems of the World. Study Committee FID/RI. FID Publ. 659. Moscow, 1987. 95–104.p.

THE "AWAKENING" OF SCIENTIFIC INFORMATION
(Social Demands in Information)

For centuries the relationships between science and special literature, socio-economic development and the gathering of information remained virtually undisturbed; there was an obvious division of labour and equilibrium between them. Without any complicated transmissions and information services, scientists gathered information, partly directly from one another through correspondence or meetings, and partly by using library collections.

Special Literature and Milk Chocolate

Even in the second half of the nineteenth century, when the above-mentioned equilibrium between research and information gathering began to tip over, inventions of great significance came into being in such a way that merited no information in special literature. Milk chocolate, for example, one of the greatest innovations of the food industry, which is probably one of the most popular food products to this day, was made without any traceable impact in special literature. This product of world-wide popularity was worked out experimentally by Daniel Peter in 1875 in Vevey, Switzerland, although at that time there were about one thousand learned journals published and the Chemisches Zentralblatt (1830), the forerunner of abstracting journals, had been known for several decades.

Over the nearly three centuries following the publication of the first scientific periodicals in the seventeenth century (i.e., Journal des Savants, Paris and Philosophical Transactions of the Royal Society, London) their numbers reached about 100,000. According to a survey conducted in the late 1960s, the number of scientific conferences had nearly quadrupled in 20 years, growing from 1,000 to 3,500. From the Vienna Congress to the present, more than 3,000 international organizations have been established. The great majority of them are not intergovernmental, but professional and scientific – so called NGO – associations. This picture is nearly complete if account is taken of the number of the annual volume of U.N. documents, estimated at 130,000 and filling half a billion pages in the late 1960s – but we have not mentioned the hundreds of thousands of research reports, the so called "grey literature"!

From these widely-known data, the similarly well-known conclusion is drawn to the effect that an information explosion is under way today. The data quoted above do, indeed, indicate a tendency, the essence of which is that the equilibrium between scientific work, research and the usability of special literaure has been upset. This means that the utilization of special literature could no longer be left, entirely, to scientists, researchers and librarians as their "internal", corporate affair.

"The Third Equal Partner"

Thus a new dimension emerged. The first signs governments began "to discover" the importance of science, owing to the unfolding of the scientific and technological revolution. Consequently, the socio-economic relationships of science were formulated as *national science policies* at the level of state policy. But science is, by nature intolerant of autarchy, departures from national science policies, international and regional, (then world-wide scientific cooperation – functioning within the framework of the U.N. and its specialized organs –) soon appeared. This was followed by the framing of intergovernmental and cross-national science policies. Then, with some delay, the internationalization of scientific information took place and this process is still going on today. The essence of it is that the problematique of scientific information becomes an integral part, first, of national science and then of international policy. UNISIST, initiated by the UNESCO and ICSU, and accepted by more than 80 governments in 1971, shows, among other things, that besides science, governments have "discovered" information, too.

Thus, the process of upsetting the equilibrium between research and the usability of special literature was codified, internationally, at government level. This process gained speed after World War II. Consequently, information transfer, i.e., scientific information established as transmission, itself became a speciality (information science, informatics). This process has been summed up by György Marx, physicist and member of the Hungarian Academy of Sciences as follows: "The mass of information accumulated by specialists .... should be extracted in a way that, while preserving all values, it may be suitable for human synthesis. The specialist leaning over their work-benches and the synthesists seeking to imagine the unimagined will be joined by a third equal partner – the community of information processors (Társadalmi Szemle, 1970. No. 3.).

The community of "information processors" exert however, either no or only a slight influence on keeping primary literature (books, journals and research reports) within reasonable limits, or "bringing it under regulation". In a spirit of "publish or perish", there is a nearly endless flow of publications, although some initiatives have been taken – e.g., by the Soviet academician Dubinin – towards a world-wide rationalization of the number and size of journals. The essence of Dubinin's proposal is that journals should publish only summaries made by the authors and the papers themselves should be disseminated in full by international and regional information centres (Vestnik Akademii Nauk SSSR, 1962. No. 4.). So far, these initiatives and attempts have failed to be realized, but certain international standardization measures, such as ISBN, ISSN and international bibliographical descriptions, etc. have been succesful in promoting the identification and accessibility of documents.

The efforts made to achieve a more rational utilization of secondary sources of special literature (i.e., documentation, bibliographies and abstracting services) have reached the round tables of international negotiations at government level, especially within the UNESCO framework. Among these are the international exchange of books, inter-library loans, aid to developing countries in the field of information technology, standardization of information, and the like. But these efforts are not enough for surmounting the difficulties put into words by A. A. Milne's Winnie the Pooh. Pooh goes visiting and gets stuck in a "tight place", after having stuffed himself with honey and milk. He complains in vain: "... it all comes of not having front doors big enough". He is advised to wait until he gets thin again and will be able to crawl out.

For the time being, it is not probable that primary literature or secondary sources will "get thin". There seems no other solution for us but to adapt to the existing circumstances and one basic form of such adaptation is automated documentation. This however, leads us to another topic. This sketch has sought to indicate only some special features of the process advancing from science through special literature to the awakening of scientific information.

In: Social aspects of modern informatics. Study Committee FID/RI. FID Publ. 649. Moscow, 1985. 71–74.p.

SOME QUESTIONS OF EFFECTIVENESS IN INFORMATION WORK AND THE EXPERIENCE OF INTERNATIONAL ORGANIZATIONS

Measuring effectiveness in information work in the field of social sciences and the humanities is extremely difficult. In this domain even the very concept of effectiveness is relative and its definition begs the question. In this context one is reminded of the logic of Marx's reasoning in the "Theory of Surplus Value". Marx claims that it is possible to define labour as to how much of it is necessary for making a table, meaning that there is a certain amount of work that is necessary for the production of a given object. This is impossible to do in the sphere of intellectual production. In this sphere the determination of the quantity of work required for achieving a result is as relative as the result itself.

How does effectiveness manifest itself? For a clarification of this concept one may turn to the collection of studies edited by Professor A. I. Mikhaylov "Theoretical bases of scientific information" (Russian edition 1975.).

Effectiveness manifests itself first and foremost in the profit of the area it is meant to serve. Even this is valid with qualifications only. It may occur that in a given area the information system is well-developed, yet the area the system supplies yields poor results. For instance, the system of economic information operates on a sufficient level, whereas the economy (for various reasons) tends to meet with difficulties in the supply of foodstuffs as well as in industrial production. Indeed, information by itself is unable to tackle these problems. The same applies to the international organizations. Neither the international position nor the territory of a single country may change merely by it making use of the information of international governmental or non-governmental organizations. The effectiveness of the activity of international organizations has not been substantially influenced by the level of their information work. The domain in which information is bound to be most beneficial is scientific research. It is here that effectiveness may be determined more precisely.

1. Effectiveness of information primarily manifests itself in scientific research. This reflects the positive results of the socio-economic process. How, then, could we measure – approximate it may be – the effectiveness of research work in the field of social science information? It is advisable to adopt two approaches. One is by way of interviews, the representative-demonstrative method which analyzes the users' opinion on the currently operating and the potentially feasible information systems. This may be exemplified by the following metaphorical saying: if the experienced fisherman wants to make a good catch, he will choose an alluring bait for the fish and not the one which will allure him. The other approach is to assess (in some fields) the information by advance orders, then to analyze the inquiries of subscribers to the information system. It

must be pointed out that there is no reliable indication whatever as to the real need of the consumer of information. Certainly this applies to the international organizations as well.

2. The first method, which adopts interviewing and analogue procedures, requires that the degree of use of the potentialities of the information system should be analyzed and so-called "user studies" should be periodically released containing an analysis of the users' opinions. The second method should analyze the subscribers' orders which were assumed to be the main indicator of the actual use of the system's facilities. Even more complex is the analysis of the effectiveness in information work of international organizations, because one should take into consideration their political affiliations as well. In these organizations declarations often substitute the factual measuring of effectiveness. An essential pointer of information work in international organizations is the number of their publications. This is indeed crucial in terms of the spreading and availability of the documents and publications released by them. One should refer to the vast flow of documents of United Nations and its specialized agencies in order to illustrate the significance of the bibliographic recording of documents issued by the international organizations. In addition, the meaning of the word "effectiveness" may also help to elucidate the underlying concept.

3. The primary manifestation of the effectiveness of information work in international organizations is the bibliographic registration of their publications, a provision for a retrospective search. In the field of sciences and technology it is possible to measure or qualify the effectiveness of information, since the results of research in these sciences are tangible as the new products of science and technology promote the application and utilization of knowledge. In the domain of social sciences, on the other hand, the end-product of research defies any attempt at a clear-cut definition. Although this is familiar enough, it is still worth reiterating in this context. In social science information – and this also holds true for the international organizations – one could speak of trends rather than of quantification of results. This by no means implies that the methods of quantification should be left unchanged. On the contrary, current research may bring further clarification. International science could certainly profit from a bibliographical service which is based on national literatures that yield national publications. Added to them the service can also use the best known international publications. The service can utilize the information of special literature of each country within an international framework. This already exists in terms of the series called "international bibliographies of . . ." published under the aegis of UNESCO, or the serial publications issued by the International Committee for Social Science Information and Documentation. So in international series and serials the distribution of special literature of different countries is identifiable.

4. Quantitative methods can help research in exploring the trends of growth in effectiveness partly to raise the standards of information services, partly to determine the most expedient channels for furthering information to users on a reasonably profitable basis. The access to primary documents is the main prerequisite of this profitability. To be sure, nothing can substitute for the primary documents (books, journals,

research reports etc.). The final goal of information is to make primary documents available to scientific research.

5. The fundamental basis for securing a high level of effectiveness in the utilization of information, especially in social sciences, is a well-organized and technically equipped special library which unites the information processes. In summing up I would emphasize that these conclusions reflect the work of several years. Some time ago I wrote that the most effective information service must have been in the Library of the British Museum at the time when Marx worked there on his "Capital". Yet, at that time concepts like scientific information and its effectiveness were still unknown

In: The efficiency of information work. Papers on the II. scientific conference of MISON, Tallin, 22–24 November, 1982. 118–122.p.

CLAIMS OF AN INFORMATION POLICY AND THE LIMITATIONS OF THEIR IMPLEMENTATION IN HUNGARY*

An information policy in the social and the natural sciences

There are three focal points to this argument. The first is: "is it necessary to have an information policy?", the second is (if the answer was yes to the first question): "in what kind of situations" and the third is: "what are the limitations of an information policy?". This argument represents an individual point of view and should rather be considered as a thesis for discussion than a plan for a solution. I believe that a scholarly/scientific communication has as much scholarly content as wide a discussion it induces. Scholarly/scientific agreement is hardly the norm: the same must be true to questions of principles and theory in the field of information.

But, first of all: *is there a need for an information policy?* My starting point is that Hungary is a small country as regards to its size and population yet not small in an economic sense and particularly not as regards to its place in science. According to some surveys Hungary's place is twentieth to twentyfourth as regards to its per capita gross national product.

As for the country's ranking in science the measurements were provided by the Team of Science Analysts working for the Informatics Service of the Library of the Hungarian Academy. By and large it was found to be commensurate with statistics and rankings traceable in UNESCO and UN and other similar statistics.

The scientific ranking of Hungary as twentysecond is based on the placing of scientific fields (e.g. mathematics has a "higher" rank than the rest). These are some objective criteria which should be considered when deciding whether an information policy was necessary. Added to these are the following: in Hungary more than 40 % of the gross national product devolves on economics of external trade. This open economics should be parallelled by an open technological development and an open information economy. (Not yet a policy.) In addition to these starting points one must consider that Hungary is poor in natural resources. Consequently it is vital that knowledge should yield growing dividends in the gross national product. Hungary cannot really export her natural products because there aren't enough of these. No gold or diamonds which could pay for necessities. Knowledge is Hungary's diamond. Knowledge contains information consequently it is vital that information should be an available commodity in Hungary.

---

* Based on the author's paper: "Information policy in Hungary in the context of modern technology and international cooperation." UNESCO – Computing and Automation Research Institute of the Hungarian Academy of Sciences – Technoinform Seminar, Budapest, 18 June, 1986.

Another element is the linguistic isolation. This makes the battle for information even harder. Yet the summary answer for the initial question is yes, there is a need for an information policy. In my opinion the need is already there but not the policy. There are separate approaches by the cultural organ of the government, by the Hungarian Academy of Sciences, by the National Committee for Technological Development, by the various ministries and by all those who have money and authority. I don't believe in policies, coordinations or cooperations which want to govern only by ideas and adages. Two things are needed above all: money and personal aptitude. Where these two are missing policies cannot be implemented; implementation needs will as well as means but good ideas are not enough.

The second series of questions deal with the context of information policy the *hows* and the *wherefores*. Initially we have to realise that there is no uniform model. I do not only think of UNESCO guides and similar vademecums but of possible models or examples. Neither in the literature nor in experience have I met a generally acceptable model which could be put in variable use.*

There are examples, there are general experiences applicable to particular national cases or/and cultural infrastructures but they are not concerned with information policy which Hungary could adopt. How could still be a general information policy implemented in the Hungarian context? With difficulty. The country is governed by a general political trend it has, moreover, a new economic mechanism which is built on decentralisation. Information policy should be adjoining to that but without schematic decentralisation. This is not to say that in general the new mechanism would not be right. On the contrary: from a flexible economic mechanism certain, adequate partial mechanisms should be evolved.

Interest, involvement as motivations should be applied to particular situations as befitting. "Example copying" whereby the model of the economic mechanism should be followed by cultural or information models would be of no avail. Partly, because the limitations of means and expertise require a great concentration in forming an information policy. Here we talk of concentration and not of centralisation.

Legislation is not new as regards to Hungarian libraries. But one cannot base a new information policy on old laws, or, according to current regulations. In 1966 the Economic Committee of the Government decided that each economic unit should look after its own information. These are things poles apart: library legislation which does not allow the development of an information policy and individual action whereby each unit would strive to look after its own information. Neither will bring the required results so we have to face an antagonistic situation.

On the one hand there is an economic system whose tendency is to rely on firms but, on the other hand there is an apparent system of informations which is to be organised, to be put to use. The latter cannot be divided into units because the money, the machinery and the experts and even other things are lacking. In my opinion concentration and decentralisation need to be combined in evolving an information policy

---

* Confirmed by my work in certain developing countries as commissioned by UNDP/UNESCO.

similarly to combining information needs derived from R+D with the market possibilities of the economic units. In other words: information planning ought to be combined with marketing.

In advanced capitalist societies governmental trends and economic programmes are atune to the possibilities of supplying needs and to the independence of firms (USA, W. Germany).

A policy of information should be aimed at increasing the dividends of a trade related information. What is considered "useful" information cannot be decided by the supplier only by the user. But how can the user express whether anything proved useful to him? By buying or paying a subscription to the information.

When information is not paid for it was possibly not useful in a trade related field. There is, however, another domain of information where "trade-relatedness" is not a requirement yet information should be supplied because it serves national culture. All countries which care for their own culture promote such information.

Economic usefulness is not a criterion in the humanities. Linguistic periodicals, for instance, should be getting grants, so should history, archaeology and others. They are the depositories of information that relate to national culture and form a heritage of identity. Such information is financially aided by the state partly in its own interest and partly for the sake of human resources. Information as a source of profit is incidental in such cases.

I would like to put forward the hypothesis that *the more advanced is the economy the larger is the trade related information dividend.* In Hungarian economics it is not enough an accepted fact – as yet – that *information is a commodity*; in other words it is part of the production of goods which has to be paid for like implements or natural resources. A good many of our firms are neither able to innovate nor update their technological equipment why should they be willing to pay for information? This is the chief limitation of a policy of information.

Currently the *National Technological Committee* has to deal with questions relating to information policy which means that this is not in the sphere of the cultural body which is the superorgan of libraries neither is the general policy related to the Hungarian Academy of Sciences, nor, for that matter, to the Governmental Committee of Scientific Policy.

In other words there is no specialized unit nominated by the state which should look after information policy, either on macro or on micro level.

Let us call a halt in asking questions. Nonetheless we cannot forget about the united cause of libraries the promotion of which raises doubts not to mention the fictitious unity of public libraries and information units. As a complex these are limiting factors in the context of existing legislation.

One could think of alternatives in taking the next steps. I don't wish to make a dogma of legalistic or structural initiatives but these always reflect a general viewpoint which always expresses existing interests.

But to return to one of the initial questions: how could one substitute information policies with one general policy? In trying to give an answer I have endeavoured to put forward some ideas and rephrase the question.

In: TMT, 1987.8. 363–366.p.

Appendix

A FORWARD LOOKING RESOURCE: INFORMATION POLICY AND KNOWLEDGE INDUSTRY

It is accepted internationally that the main resources of nations are: raw materials, energy and information. Here information is equated with knowledge industry, that is to say: the "production" and diffusion of such information which are necessary to the augmented reproduction of intellectual work. This knowledge industry (a part of which is industrial knowledge) – in a future dimension – is a mainstring of economics. But economic development and social progress depends on those who pull the strings. They must be equipped with the necessary information and abilities by the knowledge industry.

If information was a basic national resource then it must be treated as a category of economics. Not just information in general but a specific part of it which deals with scientific-economic-technological development and which promotes research. Information R+D belongs to the structure of the knowledge industry which makes it partly an indirect and partly a direct force of production.

In this sense one may speak of information economics which comprises the intellectual property of information (databases, surveys of subject fields, library special collections and services provided) and the hardware necessary for producing and communicating information (reprographics, computers connected with telecommunications sets). All this is an integral part of the infrastructure of creativity and science. Information economics is worked by the good husbandry of information but in its economics and social sense it is managed by an information policy. I may add that it is the case in developed industrial countries.

From the above it follows that in a country like Hungary with scientific development economic level in the middle range where half of the gross national product depends on foreign trade – which means an open economy – where raw materials and energy are hard priorities and where there is also a linguistic isolation there an information policy must be an integral part of the economic policy. And if it was not so it should be so. Yet it is not so despite the fact that in Hungarian products information should take an ever increasing share. In other words a maximal claim for information is to be expected; from an open economy there evolves an open scientific-technological development which is the basis of an open information economy. This is the process which should be governed by an information policy.

Yet there is no general information policy in Hungary and the lack of it could be a component in economic difficulties. There are numerous libraries, documentation units, money to buy foreign special literature but all this cannot work efficiently without an organized structure.

There are many reasons for that, some of which I shall list.

The network of Hungarian libraries and information services still bear the signs of extensive development and those of cultural orientation for the people, which had been evolved after the second world war and which were fully justified in that historical epoch and in some respects ever since. But the placing of information activities in government is, to this day, uncertain. (Traditionally there have been organizational frameworks on good theoretical grounds for the high level development of public libraries.)

The lack of a general overall conception meant that no information policy had been formed but there have been fragmentary information policies. These exist as fragments and independently of one another. For that reason alone it was impossible to implement worksharing (except on paper) and the dissemination of scientific knowledge based on international experience and on the Bradford Law. It's well known that the latter states that the most valuable information is created at just a few places whereas poor information is created at very many places.

In addition to that little has been done to teach information culture and information at third level. There have been some initial steps to incorporate up-to-date knowledge relating to subject literature but the overall approach is still to come. The social and material appreciation of information science bears no relation to its real importance.

Whatever is lacking is due to attitudes and in due course can be righted. What is to be done?

First of all we need to recognise that the development of information is reckoned with by governments in industrially developed countries. There are reports such as the Anderla report, prepared for the European Economic Community, the Weinberg report, prepared for the White House, or the governmental reports prepared for the Federal Republic of Germany. The commissioned reports needed a commissioner as well as direction by the state. Such direction comprises the legal regulation of the problems and questions of information.

When working out an overall information concept we need to distinguish between an information service for national culture and for the strengthening of national identity and a service for trade. The former must be funded by the government, as it is done elsewhere, the latter has to be paid for like any implements or services used by the user. One may use the working hypothesis whereby the more developed the economy the larger is its information-commodity dividend. This kind of information is all based on interest.

Up-to-date information needs new type of information experts. They are the full participants of research work and work for decision making. At the focal points of the economy, management and, decision-making and at research units such information officers are needed who sieve through secondary information sources (reference works, bibliographies) then repackage and synthetise them for desision-makers and for general research and development.

Furthermore it is important to implement the decision of a competition of the National Scientific Research Fund for an information infrastructure network and no less important to introduce information science (which includes tackling subject literature) in second and third level and in adult education curricula.

This task is far from easy but sacrifices are needed when taking the next step.

In: Magyar Nemzet (Hungarian daily), 1987.No.187. 7.p.

LIBRARIES AND INFORMATION UNITS IN A BIBLIOLOGICAL CONTEXT*

1. As a cultural establishment the library – that is to say the great library devoted to the arts and sciences – is a certain manifestation of bibliology in so far as we use this concept to mean the synthesis of information connected with writing.

There is no framework of written intellectual products measuring up to the scientific library, which contains the knowledge about everything that is connected with the written form. Such information is integrated best by the library, consequently it in itself is a topic of bibliological research, partly because it plays a role in research and development which has economic connections and significance. It is, therefore, fair to say, that bibliological research – as a synthesis of studies connected with writing – is a carrier of significant information related to cultural history. The various trends that deal with writings in libraries may be regarded as historically settled. But that already begs the question: what are the qualitative connections of the art of writing with collections of learning? The answer is manifold and parts are provided by various disciplines like: library history, the history of the book, the concept and methodology of scientific information, the history of the press and of publishing, the science of communication etc. Bibliology might aspire to connect these disciplines without taking on any of their scientific mantles. Progress in this field is thwarted because libraries are mainly concerned with the expediencies of procedural and organisational problems not to mention the questions of installing new technology of material and personnel difficulties of trying to overcome the resistance against new technology and other similar matters. Furthermore, we may also remember Marx's thought about learning, who opined that learning – the product of intellectual work – was always undervalued because the time for reproduction was not commensurate with the time that had been needed for the original production. For example, a schoolboy might learn a mathematical thesis within the hour.

Apart from their daily work librarians might profit more from doing research in another field of science or learning since that has a better standing. In libraries of humanities and social sciences the general attitudes set the tasks: cataloguing mss and old books, undertaking historical bibliography. But the extensive document production in the second half of this century made the introduction of new communication technology indispensable. These are either introduced in libraries of learning (this is mostly the case) or there is a reliance of independent information establishments. As a result, different views were formed in respect of the treatment of written documents.

---

* This is an augmented text of the paper given to the French-Hungarian Bibliological Colloquium given in the Hungarian Institute, Paris, March 1987.

2. It is the great scientific library (with its information organisation) which optimally and encyclopedically integrates written documents: manuscripts, printed sheets, books, periodicals, reprographically produced non-commercial documents, audiovisual materials, magnetic tapes and other products of informatics.

The trends and views of the treatment of written documents may be sketched thus: library science view, information centred view, historical view.

The library science view relies on the heritage of great libraries looking back at a time when the concept of scientific information work had not been formed. Specialisation was in its infancy in learning and in libraries. The library used to serve learning in its entirety in those days. Intensive work devoted to valuable collections produced a methodology used for decades resulting first in a library-centred view, and later in a view that shaped "library science" for our days. Another process, the great development of public libraries in our age took the same direction. The experiences related to the network of public libraries resulted in a methodology which homed in on library science as well.

Factually, the information centred view is rooted in economic and technological development. The Chemische Zentralblatt, the ancestor of reference periodicals started in 1830 and the rest: documents archives and information establishments followed, their rationale being their speed in processing serial literature. Keeping or conserving the documents was no longer necessary as their aim was to serve with information. This activity has peaked in the services tied to computer databases.

The historical view – which we should have taken first, perhaps – is manifested most clearly. This takes its roots in generations of librarians involved with history of the book, the printing press and the library, and it is manifested in the continuity of routine work evolved by great libraries. This is a progressive view in so far as it does not claim exclusiveness or special privileges for historical topics.

All three views are integral parts of a science and cultural policy, and in their entity they provide one of the main areas of research for bibliology.

3. In a great library of learning up-to-date communication technology is linked with functions pertaining to cultural history in such a way that is capable of displaying bibliological works harmoniously. The result may be regarded a kind of bibliological democratism.

Bibliology has significant functions as regards to cultural history whose research would necessarily be carried out in great libraries. These researches provide a framework and harmoniously combine communication technology with charting knowledge.

In this sense we may speak of a kind of "bibliological democratism". This manifests itself in being service centred as opposed to taking one of the views already mentioned. It helps to explore any and all kind of knowledge. But beyond all that, bibliology reaches further than the natural pragmatism of information services and investigates the place of writing, the producing of documents in the changes of history. In these changes it is chiefly the document that represents continuity since it is, in the main, both the form and the means of importing knowledge. The writing, the document are qualified as enemies in dark periods of history hence the obliteration of books ever since the burning of the Alexandrian library.

4. In a bibliological context the developing of a great library of learning is also an important chapter in long distance cultural planning. Collecting as a basic function must have material priority since it is one of the main insurances of cultural continuity.

In the great library the knowledge of the past is integrated with that of the present for the sake of the future. This continuity of knowledge in the great library of learning makes it necessary that cultural forward planning should have a certain "bibliological bias". In this respect there are such practical tasks which provide for the developing of a network of learned libraries with the help of such theoretical research which includes bibliology as well. It goes without saying that adequate financial help is indispensable. Thus bibliology and practice join hands.

Practice, equipped with the fruits of bibliology as well, must know the precise role of the great library in the structure of culture and it ought to know how to prognosticate its efficacy. The expected effect cannot be divorced from the planning of the future when, according to Marx, free time will be the hallmark of economy. According to him the function of free time is whereby man can develop his faculties manifold. This "épanouissement" rejuvenates the renaissance ideal of man, (as opposed to narrow specialisation) in terms of the masses. Beside his activities in the production process, man can cultivate himself to an extent that it positively affects his own force of production. Bibliological researches can contribute to this expected historical position.

In press. To be published in No. 27. Revue Internationale de Bibliologie, Paris

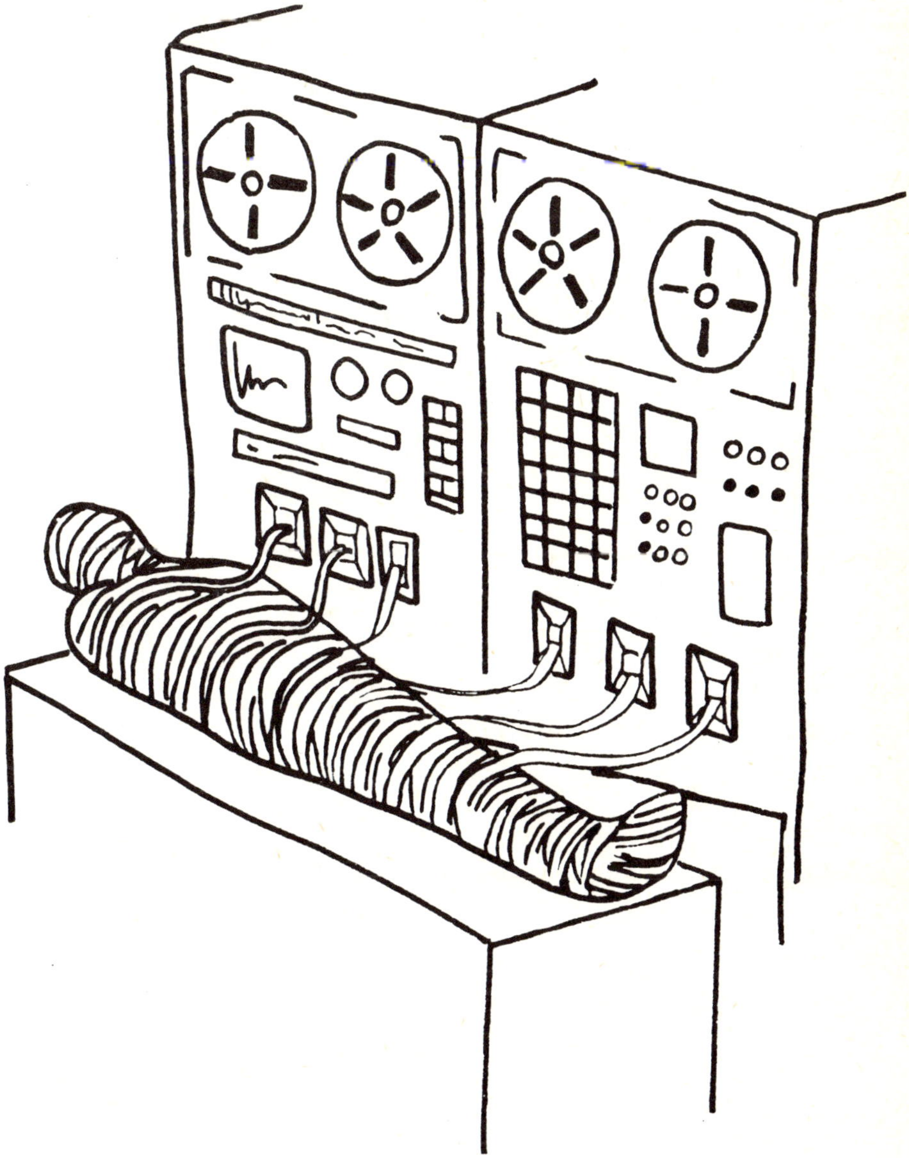

II. INTERNATIONAL RELATIONS IN THE FIELD OF SCIENTIFIC INFORMATION

UNITED NATIONS LIBRARY AT GENEVA: AN INTERNATIONAL RELATIONS RESEARCH CENTRE

The Main Characteristics of the Library

The United Nations Library at Geneva occupies a very particular place in the U.N.-family of libraries and documentation centers. This is due to its *historical character* (it was founded with the League of Nations in 1919); to the richness and the variety of its holdings (it is at the same time a library, an archive collection and a museum); to the diversity of its functions, due to the wide variety of organizations and individuals served; to the amalgamation of its official and academic tasks (it is the depository institution of the complete sets of the League of Nations and U.N. documents and of important collections of academic publications); to its size (it is the largest library of the U.N. system and one of the most important on a world-wide scale in international affairs); to its special function in relation to Swiss libraries in particular and *European libraries* in general (for these geographic areas it is the basic library for international organization documents); to its role of a reference-centre for international meetings (Geneva is the largest centre of IGO's and intergovernmental meetings – beside the U.N. meetings, the European security conference, the SALT, the Middle East Conference, etc.); to the very close links with its principal users (the parent organization and member governments are its main suppliers of documents and at the same time the main users of its services. This, however, is a general characteristic of the IGO libraries).

To some extent the Library with its collections and functions reflects continuity and change in world affairs in a special way. The events and the history of the last half century are condensed in its collections.

The Background of the Library

The Library was set up in London (at Picadilly) in 1919 with the League of Nations. It was transferred with the Secretariat of the League of Nations to Geneva in 1920. The first headquarters of the League of Nations was in the Hotel National until 1936 when it was removed to the Palais des Nations where the Library occupied its own wing.

In 1927, the League of Nations was offered a grant of $ 2 000 000 by John D. Rockefeller Jr for a library building and endowment fund with the aim of the Library "serving as a center of international research and an instrument of international understanding". The dual aspect of the library, on the one hand, to act as a secretariat library (including services for conferences, delegates, journalists), on the other hand, to act as a research centre, has remained the special feature of the institution. The library policy was formulated by a special committee of the League of Nations comprising political figures, researchers and librarians under the chairmanship of the international jurist Vittorio Scialoja. Since then growth of the library has never stopped. Even during the Second World War in spite of budgetary, staff and other difficulties, it was possible to maintain its collections, although on a limited scale. After the war and in 1946 when the library was taken over by United Nations, it was even then able to provide considerable research facilities. It is easy to understand the significance of these facilities for governments, for the newly established international organizations and for scholars in ruined Europe.

The main strength of the library activities has always been its close links with the parent organization (League of Nations then United Nations) and the governments, the principal suppliers of its collections and their most important users. This homogeneous and multi-faceted process is illustrated below:

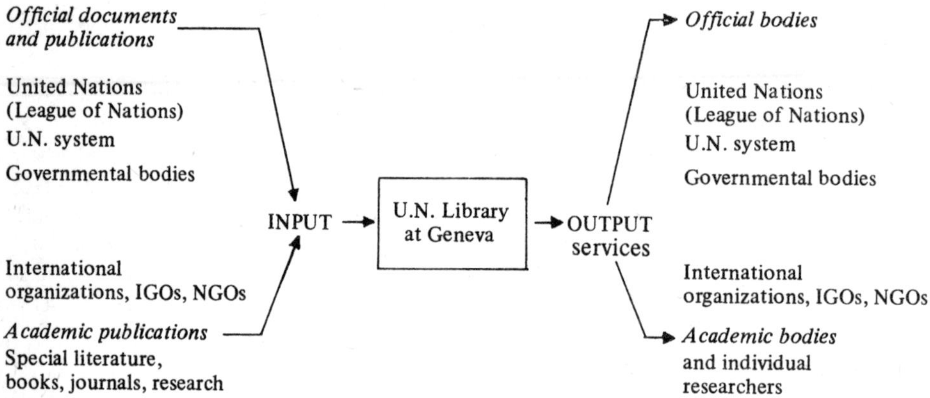

Thirty-five Kilometres of Shelves

The number of volumes bound is about 800 000. Current serial titles received are about 10 000. The number of documents can be counted in millions. The real estimate of the size of the collections of the Library is the volume of its shelves. It is of an order of about 35 km (10 floors of iron stacks and the shelves in the reading rooms). The increase of the holdings per year is approximately 13 000–15 000 volumes. The most important collections are the following.

United Nations and League of Nations publications and documents

These are the most comprehensive collections in the world including all documents in all official languages of the Organization, also comprising several thousand microfiches.

League of Nations Archives and other historical archives

The League of Nations Archives are a unique collection in the world for research in contemporary history together with several important other collections and funds such as of the Bureau International de la Paix, of the Suttner-Fried, of Lyautey, etc., funds to which are to be added the Museum's collections (paintings, rare vases, sculptures, caricatures, photographs, works of art and several historical items of the period of the League of Nations).

U.N.-system publications

An important collection of the U.N. specialized agencies publications and main documents.

International organizations publications

The publications and the main documents of the IGOs (and of some NGOs) such as CMEA, OECD, European Communities, etc.

Governmental publications and documents

An exhaustive collection of official gazettes, parliamentary debates, statistics, national plans, governmental reports and other official papers.

Legal materials

"... there is no law library as such in the Palais des Nations, a basic reference collection in public and private international law, comparative law, civil, procedural, constitutional, criminal and commercial law. . . ."[1]

Publications in economics

Data, statistics, tariffs and taxes, yearbooks, economic conditions of regions and individual countries, economic development, problems of the developing countries, trade policy, economic integration, monetary systems, public finance, industrial and energy problems, etc.

International relations, politics

Theory of international relations, representative monographs on policy of individual countries and regions, diplomacy, developing countries, security, disarmament, apartheid, etc.

Social questions

Social progress, condition of women and of youth, narcotic drugs, refugees, disaster relief, etc.

General works

Data, directories, encyclopaedias, dictionaries, terminologies, guides, "Who's Who".

Library and information works

Printed catalogues of books, union catalogues of periodicals, national and special bibliographies, abstracts, indexes.

In addition to the above types of publications and documents, the library acquires on a continuous basis books and periodicals reflecting the particular interests of the organization, e.g. application of science and technology, contemporary history (especially the inter-war period and World War II), transport (aviation, shipping).

Services

The library is open to staff members of the U.N. secretariat, the permanent missions at Geneva, the delegates, the accredited journalists and to all researchers who can justify the need to consult the U.N. Geneva Library's collections.[2]

Consultation of the library's holdings is carried out in the General Reference Unit or in the specialized reading rooms (about 120 places). U.N. documents, legal and political, economics and transport, international organizations' social problems; the specialized reading rooms are equipped with a selection of books and periodicals in their respective fields, a self-list and card-index of selected periodical articles, this latter being published in the Monthly List of Selected Articles.

Outside loans from the Palais are only exceptionally granted. The non-official readers (outside the U.N. and governmental delegates) are supplied with reader cards, the number of these per year is about 1000. Most of them are scholars from all over the world (professors on their sabbatical year, Geneva and French university teachers, study grant holders, etc.).

The U.N. bodies served by the library (without the specialized agencies located at Geneva which use its services to a limited extent) are the following: Economic Commission for Europe (ECE), UNCTAD, High Commissioner for Refugees, Narcotic Drugs Division, Opium Board, Disaster Relief Bureau, Division of Human Rights, CESI (Centre of Economic and Social Information), the services related to UNDP, to the Environmental Program, to UNICEF, to UNITAR, the IOB (Inter-organization Board for information systems and related activities), the liaison bureaux of different U.N. bodies and departments outside Geneva such as Inter-Agency Affairs, the Geneva Branch of the Office of Science and Technology, the common service of ECE/FAO and the U.N. Office at Geneva itself (the Legal Adviser, the Linguistic Division, the Office of Public Information, the Conference Services, etc.). totaling about 3000 officials. Some economic organizations such as the GATT and the International Trade Centre also use the Library to back up their own special but limited collections.

In 1974 some 5600 meetings were held in the Palais. In Spring 1975, the Conference of the Law of the Sea was attended by more than 2000 delegates who made 20 000 pages of photocopies from the collections of the Library. The number of personal consultations in the special reading room organized for the delegates cannot be counted.

In addition, several important inter-governmental conferences and meetings are held outside the Palais, e.g. the European Security Conference for which the Library is also open.

Counting about 1000 non-official readers, 3000 U.N. officials, about 5–5500 meetings per year in the Palais and the organizations and the meetings outside the Palais which also use the library, all these give a picture of the variety and volume of services provided by the library. Some reference and documentation units serve as transmission centres between the library and the division located in the Palais. The major information tools of the library are traditional. The main catalogue (a cross-catalogue of authors, titles and subjects) is based on the Anglo-American Cataloguing Rules. For classification the Library of Congress subject headings and the U.D.C. (Universal Decimal Classification) are used. An automation project has been launched with a view to computerization of the catalogue of periodicals.

Publications and Activities Outside the Organization

The two main publications of the library have been launched in the period of the League of Nations: the *Monthly List of Books Catalogued* (MLBC) in 1926, the *Monthly List of Selected Articles* (MLSA) in 1927. Together, they contain about 13 000 entries per year. Both have a world-wide diffusion with several hundred subscriptions. The MLSA is published in about 1800 copies and this periodical together with the MLBC serves as the basic exchange of publications material of the library. This activity is a very important one: approximately 76–80 % of the acquisitions of the library are acquired through this channel. The titles of the articles in the MLSA are selected from more than 3000 periodicals. The cumulation of selected chapters of the MLSA has been published in several volumes by the Oceana Publishing Co., New York. For the last two years, the MLBC has contained, in addition to the usual bibliographical details, a descriptive subject-analysis aimed at a future automation programme. Furthermore as a by-product of the MLBC, printed cards are diffused for the benefit of the U.N. depository libraries.

For quick information about the new books for the divisions in the Palais, a Daily List is distributed upon request.

A *Catalogue of Periodicals*, annuals and special series currently received in the U.N. Library was published in 1972 (in 454 pages) containing more than 10 000 titles with their subject analysis. It is the most important catalogue of periodicals in the U.N. system.

The series of Reference Lists was launched in 1970 based upon the needs of the organization in co-operation with the operational divisions. Several bibliographies and indexes were published (in 1200–2000 copies) such as *East-West Trade* (No. 1, 80 pages), *International Law Commission, a Guide to the Documents, 1949–1969* (No. 2, 55 pages), *Science Policy in ECE Countries* (No. 4, 68 pages), etc. A special mention should be made for the index containing about 8000 of references to U.N. documents prepared for the Middle East Conference held in Geneva at the end of

1973, then published in microcard form with the title *United Nations Documents on the Situation in the Middle East, 1947–1973* No. 7, 7 microcards).

Beside the above publications, the Library has largely contributed to the publication of two basic volumes on the documentation of international organizations: *Sources, Organization, Utilization of International Documentation. Proceedings of the International Symposium on the Documentation of the United Nations and Other Intergovernmental Organizations held in Geneva, 21–23 August 1972.* The Hague, 1974, 586 pages, FID Publ. 506., and *Documents of International Organizations: a Bibliographic Handbook* covering the United Nations and other intergovernmental organizations. Compiled and edited by Th. Dimitrov. London, 1973, International University Publisher, 301 pages.

In fact, these two volumes were the result of many years' joint efforts by several professional organizations and the Geneva Library in launching training programmes and in developing the appropriate methods in handling documentation of international organizations. As Professor F. Casadio the Rapporteur general of the Geneva Symposium pointed out, summarizing the recommendations of the symposium, both international organizations and governments must pay more attention to the documentation of international organizations if they want this documentation to be usable. The library has organized in co-operation with UNITAR and other bodies seminars, training courses, etc., which were attended by about 150 diplomats, documentalists and government officials from all over the world.

In addition, the library staff members have participated in several seminars and colloquia to promote international co-operation in documentation distribution and treatment.

About Historical Collections and League of Nations Archives

It would be worth while to publish a special study on the historical collections and League of Nations Archives. The archives were transferred to the library in 1958, but only since 1965 was it possible to convert the files of the Central Registry and the documents and manuscripts of the League of Nations into a real archive collection when the Carnegie Foundation for International Peace offered a grant for this purpose.

The rule of access to League of Nations Archives has been signed by the Secretary-General of the United Nations, U Thant, at the end of 1969 (26 December, ST/SGB/135). The major parts of the historical collections are the following: *Archives de la SdN 1919–1947, Archives du Bureau International de la Paix 1889–1950, Fonds Suttner-Fried 1885–1921, Fonds divers 1908–1952.*

The following information tools were established to facilitate research in the historical collections: *Guide to the Archives of the League of Nations* (also in a French version) and the *Archives de la Société des Nations, Réportaire général 1919–1946,* Vols 1–3.

By a special budget allocation it was possible to organize a laboratory of de-acidification for the conservation of historical collections documents. The conservation programme will take about 20 years. A programme of microfilming all archives documents is to be established in the near future. It is hoped that a League of Nations Archives publications programme will be realized, covering registry and section files, records, etc.

Problems and Prospects

Like any institution the library has its problems. In the overall framework of the U.N. there is a need for a well thought-out system of co-operation with dispersed divisional documentation and reference units. An ambitious but realistic automation programme should be envisaged for research purposes. The budget is still far too limited, as is the number of staff, (all told, 45). With a healthier budget and an improved personnel situation, the library would be able to extend and ameliorate the service it can render the U.N., other international organizations, governments and the world community of scholars.

NOTES

1. B. I. Ross: (Dec. 1973 – Mar 1974). The Legal Section of the United Nations Geneva Library. *Law Librarian (London)*, 4, (3). This article gives a brief and very good picture of the Library's activities in legal field, on its collections and services.

2. The Library acts as a division of the U.N.O.G. under the direction of the Chief Librarian; the supervisor of the library is the Director-General of the Geneva Office. The structure of the library is the following: Office of the Chief Librarian, Processing Section. Readers' Services and Documentation Section, Historical Collections and League of Nations Archives.

BIBLIOGRAPHY

United Nations (1966). *The Libraries of the U.N. A descriptive guide.* 126 pages. New York.

Gardé, P. K. (1970). *The U.N. Family of Libraries.* 252 pages. New York: Asia Publishing House.

Dale, D. C. (1970). A chapter and a bibliography on the Geneva Library is to be found in *The United Nations Library. Its Origin and Development.* Chicago: American Library Association.

Ghébali, V. Y. (1973). *A Repertoire of LoN Serial Documents, 1919–1947.* Vols 1–2. Oceana Publishing.

Marulli, L. (1979). *Documentation of the United Nations system.* 225 pages. Metuchen, N. J. – London: Scarecrow Pr.

Dimitrov, Th. (Comp. and ed.) (1982). *International organizations. United Nations.* 41–119. pages. Berlin: de Gruyter.

In: International Library Review, 1976.2. 119–126.p.

SPECIALIZATION AND INTEGRATION: SOME ASPECTS OF THE DOCUMENTATION OF INTERNATIONAL ORGANIZATIONS

Introduction: Autarchy or Integration of Efforts and Knowledge?

The role played in progress by the *division of labour* is well known. The *scientific and technical revolution* accentuates this process. The division of labour is characterized, among other traits, by *intensified specialization*, whose results are well known: the proliferation of branches of human activity, the creation of new disciplines, and a tendency for a branch of science which had long remained a unity to divide into several branches. The consequence of this process is the division, diversification and ramification of *specialized literature* and, above all, of periodical journals and series. This process lies at the root of the problem expressed in the concept of documentation, but the phenomena of the division of labour and of specialization appear also in the area of *international relations*.

The many changes that have taken place in the *political geography* of the world since the Second World War, the emergence of a whole host of States who have recently achieved independence and, side by side of the increasing diversification of international organizations and their activities have given rise to a specific documentation problem: that of *international documentation.*

Faced with this complex problem of specialization and of the development of international relations, the international community believes that *autarchy is impracticable* and that the appropriate solutions to the different political, economic, scientific and other questions should be sought in the possibilities of multilateral *institutional* collaboration and co-operation offered by the international organizations.

From the institutional standpoint, therefore, the international organizations represent a *counter-trend* towards integration, as against the specialized efforts of the different States, professions and branches of human activity.

1. The Growth Trend of International Organizations

In international organizations the general phenomenon of specialization reflects, from the *institutional* standpoint both political considerations, and the specialization of science, technology, production and institutions. This is evident from a comparison of the League of Nations system with that of the United Nations. The relatively *homogeneous* institutional character of the *League of Nations* has developed into the multiplicity of bodies of the *United Nations system* (the United Nations and the specialized agencies). This transformation may be regarded as one of the most important *consequences* of the political, economic, social scientific *development* that has taken place since the Second World War, and exhibits a tendency towards *internationalization* which is also evident in the specialization of non-governmental organizations (NGO's) and in the multitude of regional intergovernmental organizations. A publication of the Union of International Associations, *The 1,978 International Organizations founded since the Congress of Vienna, Chronological list,* Brussels, 1957, lists the international organizations, associations, unions, etc. (intergovernmental and non-governmental) from the Congress of Vienna to 1957, all 1,978 of them. A statistical table giving the number of international associations founded in each five-year period shows that the greatest number were founded in the period 1950–1954, i.e. 319 NGO's and 33 intergovernmental organizations. There were *2,500 international organizations* in existence in 1971.

International co-operation, then, manifests itself through specialization both in institutions and in intellectual output: publications, specialized literature and documents. Only recently the Secretary-General of the United Nations affirmed that *the Organization's most important working tools were documents.* Thus the main medium for conveying information consists of documents.

The world "information" is used here to mean all information and data which, in content and presentation, may help to improve and extend knowledge and the administrative, economic and scientific activities dealt with by international documentation services.

As in science, a process of integration and synthesis is taking place which aims at generalizing the results of specialization. Integration of information therefore complements specialization and the two together form a unity in the division of labour.

In the words of Mr. René Maheu, Director-General of UNESCO, in *Science et Société*, Paris 1967: "... Faced with the growing specialization of both thought and action through the diversification of research and the division of labour ... man ... is as likely to be choked by his knowledge as paralyzed by his ignorance."

2. The Growth Trend of the Volume of Publications of International Organizations

The growth trend affecting the volume of journals is evident from the *Auger report* published by the United Nations and UNESCO in 1961:

beginning of the 19th century:	100 journals
1850	1,000 "
1900	10,000 "
1960	nearly 100,000 "

If this growth rate becomes constant, there will be some 1 million journals in circulation by the turn of the century. According to estimates made by Allen KENT in an article entitled "Resolution of the literature crises in the decade 1961–1970" (*Research Management*, 1962, No. 1, pp. 49–58), about 2,000 books, newspapers, reviews, reports and other publications and documents, with a total volume of some 1,050 million pages, appear every minute throughout the twenty-four hours of a day.

What is to be said about the most noteworthy production of all in terms of both volume and variety of subjects dealth with in *United Nations documents*? Figures – not estimates – show that a truly impressive contribution comes from the source. The Secretary-General's note on the budget estimates for the financial year 1970 concerning the production *of documents* (twenty-fourth session of the General Assembly, document A/7576 of 25 July 1969, para. 2) shows that, in round figures, the growth of the documentation produced internally at Headquarters was as follows:

1964	400 million page-units
1966	500 " " "
1967	600 " " "

In percentage terms, this means an increase of 50 per cent for the period 1964–1967.

For the forty-fifth session of the Economic and Social Council, 130 documents were reproduced in New York, Geneva and elsewhere; a complete set of these documents alone weighed 18 kg and formed a pile 50 cm high (op. cit., para. 159). As a matter of interest it may be mentioned that the delegation which asked for the greatest number of copies of each document – 145 – received a mass of paper weighing 2.5 tons and measuring 75 metres in height.

Even if we leave aside the fact that the *composition* of these 600 million page-units varies very widely as to *content, length, reproduction figure, language* and *variations on the same theme* – ranging from bulky documents of great political and economic importance, in three languages and a large number of copies, to short administrative circulars of which only a few copies are made in one or two languages – it must still be borne in mind that the figure of 600 million pages represents only mimeographed and offset production at Headquarters – no printed items – and furthermore, that it does not include the output of the Geneva Office and the specialized agencies.

It is thus quite understandable that steps have been taken to *reduce* the volume of documentation produced by the United Nations. On this subject, see the instructions given in General Assembly resolution 2836 (XXVI) on publications and documentation of the United Nations, which was adopted on 17 December 1971. It is not merely a question of taking economic action bot of avoiding, or rather minimizing, the danger of being "choked" by the avalanche of documents.

The proposals made by Mr. Hubert CURIEN, Director General of the national Centre for Scientific Research (*Centre national de la recherche scientifique* – CNRS) and published in *Le Monde* of 6 January 1972, according to which "brains must be recycled to prevent their sterilizing pollution by the over-accumulation of information ...", are not perhaps applicable to science alone.

3. Some Comments on the Terminology of International Documentation

Information is dependent on documents, which in turn form the main part of libraries, especially *international libraries*.

What is meant here by the term "international libraries" are the libraries of inter- and non-governmental organizations and their respective documentation services.

To be more precise, the notion "specialized" should be added, for every international library is a *specialized library*, even the biggest ones with the most varied collections. In short, we are concerned with a network of libraries and of specialized international documentation services.

The next point to note is that the world *"documents"* is used here to mean *official productions* of international organizations, Governments and governmental bodies, whether in the form of mimeographed documents or in that of official publications for sale.

As to the notion *"international"*, it is obvious that any documentation service and any scientific library whatsoever is international in its collections; science knows no frontiers, so no country and no organization can adopt a system of *autarchy*.

In this sense of the world, *all* scientific and specialized libraries and documentation services are international; this is the objective factor in the nature of the *collection of documentation*.

From the institutional standpoint, however, only the documentation services and libraries of international organizations can be regarded as international.

Where both factors – collection and institution – happen to *coincide*, we are faced with a *special type* of library and documentation service, *that of international scientific information services*. Since such services belong to international organizations, they share the organizations' tasks and responsibilities to some extent.

It will be well to identify two aspects of those responsibilities which are mainly of a professional nature. In the first place an international service, while retaining its scientific initiative, is closely bound up with the requirements of the organization it serves and, in the last analysis, with the requirements of the member Governments: that is, *of the international community*. Secondly international documentation services must as far as possible base their information work on document and *on official publications*, both national and international, especially where the information relates to a particular region. This *spirit of objectivity* in international documentation services is an institutional characteristic of theirs and it is precisely this *international institutional character which is specific* and which distinguishes this type of international service from other professional types.

First of all, international documentation services may be distinguished from one another according to their orientation in serving the main group of *users*. In order of priority, the first type is oriented towards the *international organization* itself, in other words, *the Governments* themselves, and the second towards the *scientific community*. These orientations must be harmonized.

International documentation services are not only responsible to international organizations and Governments but, without forgetting the order of priority, they are also responsible to the world of research. Such services therefore *occupy a place* in the world community of scientific information services while retaining their *specific character*.

In addition to official publications, international documentation services draw upon specialized literature – books, series, journals and other scientific publications – without which they would be no more than repositories, mere catalogues of official publications, mere bones without flesh. It should be noted that, despite its powers and specific official functions, international documentation is no island in the midst of a sea of information.

4. Aspects of the Orientation of International Documentation and its Sources of Information

With regard to the *utilization aspects* of international documentation; emphasis must be placed first of all on the *possibilities* offered by the documents of international organizations for purposes of the *economic, scientific and technological activity* of States. The ambitious programmes of the United Nations system for a scientific decade, such as hydrology, the development of short-, medium- and long-term scientific projects and activities, the Geophysical Year, the desalinization of seawater, or again the programme for arid zones, etc., all afford useful information – leaving aside foreign policy – on *the economic and scientific policy and the technological progress of States*. A long-term programme (hydrological, meteorological or soil improvement) may at the same time furnish *suggestions* for industry, foreign trade and technological progress. *Today's resolution* instituting a long-term project contains the elements of production and export, even intellectual export, of the *day after tomorrow*. Consequently, the processing and abstracting of the documents of international organizations *go well beyond* the normal processing of publications in libraries. Such documentation calls for *systematic, dynamic and highly specialized analysis*. The documentation services for international organizations can facilitate this work by producing an up-to-date *index of their documents*. What is needed is a *constructive co-operation* among international and national information services, including the *transmission* of "national" information to international organizations.

Specialized literature as a source of information for international organizations has two aspects which may be mentioned: firstly, information *on* the activities of international organizations and, secondly, information drawn from the various political, economic, scientific and other activities of the international organizations which are used as *sources for* the preparation of the organizations' documents. There is a *mutual influence*, sometimes direct and sometimes indirect, between the documents of international and national organizations and those of specialized literature. This *interaction* should therefore be just as valid for the respective documentation services, though it must be pursued more deliberately and directly, as it is for official documents and publications.

5. Some Conclusions: Specialization and Integration of International Documentation Services: an Information Network

There are certain conclusions to be drawn concerning the relationship between information and the documentation activities of international organizations.

Firstly, given the increase in the number of organizations and in the volume of their publications, official and unofficial, collaboration between the documentation services of those organizations must be considered essential. The notion of a *network* of documentation *services* concerned with international affairs is steadily gaining ground. This network should be regarded as a *unity* of action, as a network potentially in existence already or as an *integrated system of the division of labour*.

The term "*network*" should be understood to mean, not an *administrative unit*, but rather an *integrated scientific system* based on *principles* derived partly from the *specific tasks* performed *by international organizations* and partly from the *theory* and *experience of scientific documentation*. On the basis of such principles it would be possible to formulate a *general conception* of *co-ordination* and *co-operation* between international documentation services with due allowance for the *independence* and *specific interests* of each service participating in the network.

What action should be taken, how are co-ordination to be developed? What might be the specific tasks and methods of the intergovernmental and non-governmental documentation services?

It would, of course, be possible to draw up a list of the questions which need answering but that is not the purpose of this study, for we feel that it is as easy to enumerate our common tasks as it is difficult to find suitable practical solutions. Simply by way of example, however, the following may be mentioned: the acquisition of official publications or official gazettes from a library that would agree to their reproduction and circulation on microfiches for the network and the regular exchange of bibliographies and unpublished documents through a sort of clearing-house.

This study is inspired solely by the desire to contribute some ideas to the formulation of a general concept of co-operation between international documentation services and at the same time to raise certain questions concerning the place of these services in the world system of scientific information services.

Without leading to specific solutions, the very formulation of the problems to be discussed may, however, justify the adoption of a certain theoretical approach. As Shakespeare put it in the Merchant of Venice: "If to do were as easy as to know what were good to do, chapels had been churches and poor men's cottages princes' palaces".

* * *

See also: some French versions of the author's study in international publications:

(1) La spécialisation et l'intégration: quelques aspects du travail d'information des bibliothèques internationales, symposium de l'Association des bibliothèques internationales (A.I.L.) Vienne, Août 1970. Conférence.10 p.

(2) La spécialisation et l'intégration: quelques aspects de la documentation des organisations internationales. International Associations – Associations Internationales. (Brussels) 1972. no. 10. 452–459. p.

(3) La spécialisation et l'intégration: quelques aspects de la documentation des organisations internationales. Società Italiana per la Organizzazione Internazionale. Roma. Sezione di Milano. 22.01.1973. Conference. 11 p.

* * *

In: Sources, organization, utilization of international documentation. Symposium, Geneva, 21–23 August 1972. FID Publ. 506. The Hague, 1974. 20–27.p.

RELATIONS BETWEEN INTERNATIONAL DOCUMENTATION SERVICES

a) . . . according to orientation: (institutional element)

b) . . . according to sources:

c) . . . according to information on and for international organizations:

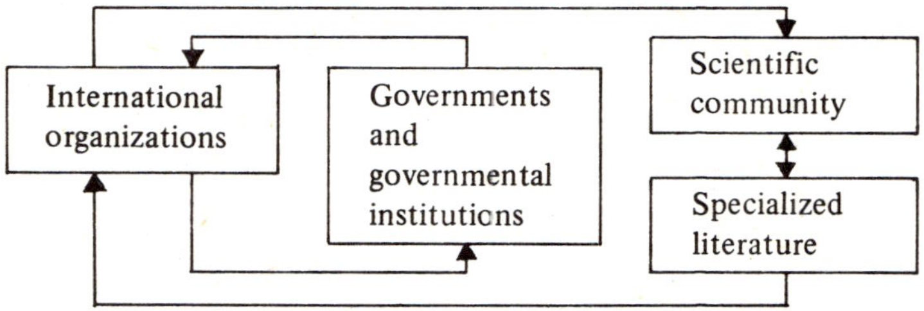

d) . . . according to priorities:

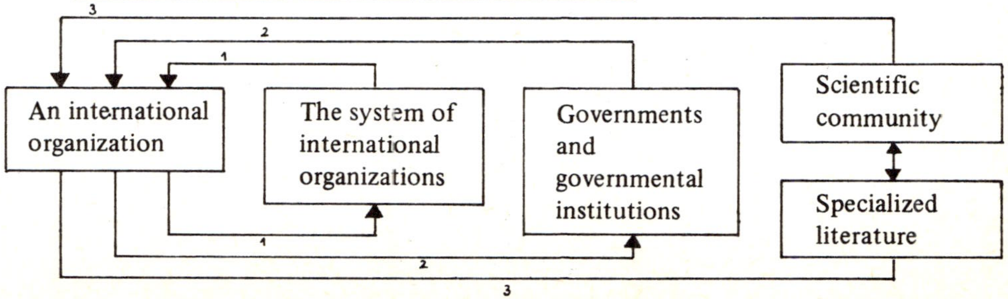

INTERNATIONAL CO-OPERATION AND TRENDS IN SOCIAL SCIENCE INFORMATION TRANSFER

Co-author: Tamás Földi

Introduction

One of the most important discoveries of the last decade – significant beyond science itself – is the perception that it may be fatal for mankind to lose control of the negative consequences of the overall scientific and technical development, e.g. the impending exhaustion of energy sources and raw materials, overpopulation, the harm of urbanization, pollution of the environment, acute food shortages, and the ever increasing and costly production of nuclear weapons and missiles. All these are so-called global questions. The natural and technological sciences can merely offer a partial answer, providing as it were the means only to master these problems, the solutions of which are essentially of a social character. This perception has also contributed to the increasing significance of the social sciences and their proponents, and partly explains the increasing interest in social science information.

The inter– and/or multidisciplinary character of the aforementioned global questions, including the new economic order, technical assistance to developing countries, etc., points to a complexity requiring increasingly more information.

In several countries, e.g. in France and Hungary among others,[1] special teams have been set up to develop and co-ordinate social science information.[2]

Information transfer is particularly significant for developing countries, having small populations and languages outside the linguistic main-stream; theirs is normally a flexible economic, scientific and cultural framework. These countries cannot be self-sufficient in the field of information; rather they must consciously utilize the widening possibilities emerging from the transfer of information. But this article is not concerned with oldest and, at the same time, the most permanent form of information transfer, i.e. immediate, written and oral communication between scholars.

Traditions of the international information transfer can be traced back to the exchange of publications by libraries as early as the 18th century,[3] to international inter-library lending, to the reciprocal supplying of bibliographic information, and to co-operation in international abstracting and indexing of journals.[4]

So the current information transfer is not unprecedented yet it has rather a new quality, with a greater dimension of co-operation than before.

There are national initiatives underlying the global scientific and social questions. These are aimed at co-ordinating national and international social science information channels. There are political considerations of détente, too, a concept borrowed from international politics and here applied to international information transfer.

Information transfer in the social sciences

Role and mechanism

The status, role and significance of information in general is closely interrelated with the characteristics of a given discipline. Social sciences develop under the influence of different ideologies which in turn reflect various economic, political, social and cultural approaches to relations among countries.[5]

Different value systems make different demands on the social sciences thereby limiting the intensity of information transfer. However, numerous countries of different social structures share important general problems in the realm of what U Thant proclaimed the three Ds: decolonisation, development, disarmament. Social science information transfer is but one example of shared concerns in an area of peaceful coexistence and widening international division of labour.

An important characteristic of most of the social sciences is the significance of written communication representing both the source and product of research.

In view of the fact that the methods of social science information transfer has for a long time remained unchanged, the recent use of computers for such transfer has aroused international interest.

In order to understand the mechanism of information transfer better, it should first be stated that over 100,000 volumes of monographs and some 5,000 to 6,000 journals are published yearly in the field of the social sciences. In addition, research reports, reports of banks and large enterprises – the so-called grey literature, not usually commercially available – and results of statistical, sociological and other surveys, increasingly play an important role in information transfer. This enormous quantity of data in social sciences needs to be processed. This processing must take into account the fact that social science information, unlike information in the technical and natural sciences, is not only aimed at experts but at everyone. That is to say, social science information must meet the demands of research, management, medium-level specialists, educators, mass media, etc.

Institutional structures

The processing of this complex social science information is promoted by a network of institutional information systems structured along the lines of international co-operation. Mention should be made of various systems which promote specific social science information transfer, such as EUDISED in the field of education, the Documentationsring sponsored by the Rationalisierung Kuratorium der Wirtschaft at Frankfurt am Main (Federal Republic of Germany), in the field of management, and the information services of the International Statistical Institute and the International Association of Law Librarians respectively.

Regional, political and economic co-operation also points to an increasing transfer of social science information. For example, in 1976 the academies of sciences of the

European socialist countries and Mongolia[6] concluded an agreement on the elaboration of an international system of information in social sciences known as MISON.[7] Similar co-operation in the sphere of social science information exists among the countries of the Common Market, the OECD, CMEA, as well as among Arab countries. At the international level the Data Retrieval System for the Social Sciences (DARE) and the International Information System of Social Science Concepts (INTERCONCEPT) of Unesco[8] are of particular importance.

The significance of the role of international organizations in the field of information transfer lies above all in their promoting favourable conditions for activities in this field. These organizations are instrumental in promoting various meetings among information experts from different countries, and in establishing personal partnership between them. The international organizations help set the course of national information policies as well as guide the attitude and interest of experts in the direction of international interconnections for information transfer. Through standardization these organizations also play an important role in bringing the information technologies closer together, thereby paving the way for favourable objective conditions of international co-operation and information transfer.

While the activities of international intergovernmental organizations are very useful from many points of view, the effectiveness of these activities is limited. Most of these organizations are more competent to discuss theoretical information policy issues than to implement concrete information programmes. The effectiveness of the intergovernmental organizations' work is often impeded by complex political and administrative considerations. Moreover, such effectiveness is primarily contingent on the financial means rather than the legal status of these organizations. This is especially true for international organizations instrumental in the transfer of social science information. By contrast, the international non-governmental organizations generally have limited resources and staff – frequently working on a voluntary basis – but because of their direct links to information services, their expertise is still indispensable in promoting international co-operation.

The following survey examines selected, significant, mostly international institutions and organizations which promote the transfer of social science information.

ICSSID[9]

The International Committee for Social Science Information and Documentation is a non-governmental international organization which was established in 1950 with the support of Unesco. The activity of ICSSID, carried out at its own initiative or in response to requests from Unesco, is concentrated in three areas: (a) elaborating proposals and plans concerning social science information policy; (b) contributing to the normative activity through the development of formats, thesauri and other means of social science information handling and processing; (c) issuing publications and thereby fostering worldwide social science information exchange; mention should be made in particular of the *International Bibliography of the Social Sciences* consisting of four series.[10] The committee has several working groups, e.g. for automation, for economic bibliographies, for political science bibliography and abstracting journal editors, etc.

FID C/3 Committee

The C/3 Committee of the international Federation for documentation was founded in 1959 by documentation experts from East and West European countries, in order to modernize the social science chapters of the Universal Decimal Classification.[11] Basically reformed, UDC Class 3 now reflects the present-day bourgeois and Marxist social science concepts. The committee's activities today embrace disciplines other than just UDC Class 3: e.g. history and management science.[12]

FID Committee for Social Sciences (FID/SD)

The committee was founded in 1978 with the purpose to provide a forum for social science experts, as well as to advise leading organs of the FID on matters of social science information development at the national and international level. The committee publishes a newsletter entitled *FID/SD Chronicle.*[13]

IFLA Social Science Libraries Section

The Social Science Libraries Section of the International Federation of Library Associations and Institutions provides social science, particularly economics, libraries with a forum for the promotion and exchange of relevant information resources in general, and bibliographic services in particular.

The section publishes social science information directories and a *Newsletter.*[14]

IASSIST

The International Association for Social Science Information Service and Technology, Ann Arbor, Michigan, was founded in 1976 in order to encourage the establishment and maintenance of social science data bases and to foster international dissemination and exchange of information connected in particular with relevant machine-readable data. The IASSIST works in close co-operation with other international social science bodies, notably the International Sociological Association. IASSIST publications include the *IASSIST Newsletter* and IASSIST Conference *Proceedings.*[15]

IFDO

The International Federation of Data Organizations for the Social Sciences was established in 1977 in Louvain-la-Neuve, Belgium. IFDO closely co-operates with ECSSID, and it was among the sponsoring organizations of the Second International Conference on Data Bases in the Humanities and Social Sciences, Madrid, June 1980.[16] IFDO activities include: data processing (especially in terms of privacy protection), standardization of data, data personnel training.

ECSSID

In addition to the aforementioned international social science information organizations, mention should also be made of the regional European Conference on Social Science Information and Documentation.[17] ECSSID was initiated by the European Co-ordination Centre for Research and Documentation in Social Sciences (the Vienna Centre), but the information transfer activities of ECSSID extend beyond Europe to North America. The first ECSSID conference was held in Moscow in 1977 and another in Blazejevko-Poznan, Poland, in 1978. Between conferences, four working groups of the International Organizing Committee of ECSSID direct policy and co-operation matters. The competence of the working groups is as follows: Working Group 1 is concerned with the exchange of primary and secondary documents on social science; Working Group 2 with the exchange of information on ongoing research; Working Group 3, the compatibility of automated information systems; and Working Group 4, education and training.

The increasing number of organs engaged in the field of social information transfer points to an intensified international co-operation in this field.

Computerized information transfer in the social sciences

Computer systems tend to intensify information transfer in general, and international social science information transfer in particular. Among such international information systems mention should be made of: Unesco's operational DARE (social sciences) and experimental SPINES (science policy);[18] in the field of labour, ISIS (Integrated Science Information System) of ILO in conjunction with other competent organizations in this field; DEVSIS (Development Science Information System),[19] under the auspices of IDRC, ILO, OECD, the United Nations, UNDP and Unesco. The significant national science automated data bases – e.g. in France, FRANCIS (social science and humanities), in the United States ERIC (education) and the Social Science Citation Index – are in fact international in coverage, thereby forming part of global information transfer.

In small and medium-size countries, one of the central dilemmas of information policy concerns the linking up with international computer systems. These countries must consider whether the traditional self-sufficiency in information practice can remain satisfactory in the future. The countries that decide to participate in international co-operation must answer still further questions. How should they choose from among existing, often parallel, systems? Is it advisable to join several systems at the same time? If so, and taking into account the frequent lack of compatibility among such systems, how should the link-up avoid straining the participants at both the input and output end? How can an institute about to link up with a national and/or international system be modified in order to meet the requirements of input tasks and output receiving? What is the optimal advisable proportion of co-operation with international information systems, co-operation which would neither cause gaps in, nor redundancy of, information. Will information systems outside the country quarantee a high quality and a rapid exchange of information? How can a given country's ideological orientation be aligned with systems in a foreign country of a different political persuasion? How will foreign political motivations influence the coverage of domestic information demands? In terms of financial outlay, is it at all economical to rely on foreign systems?

Recent information theory will hardly be able to answer these questions. The answers require extensive research and appropriate empirical data. And a pertinent answer is likely to be provided only in the course of practical operation of a given system. Therefore, it would not seem advisable for a country to plunge headlong into international co-operation in information systems and services at large; rather, it would be more expedient to co-operate initially only in certain well-defined information fields in which favourable home reception and high-quality services may be expected, i.e. areas most advantageous in terms of international, centralized information output as well as domestic capability for receiving such output and providing national (decentralized) input.

Concluding remarks

International co-operation in the field of social science information is gathering momentum. Therefore, in elaborating the concepts and development of social science information, each country should keep in mind the potential rights and responsibilities attendant upon relevant international co-operation. It is advisable to choose those international solutions which will increase the domestic feasibility of information reception and service.

It is indispensable that each country should first ensure the coverage of its own national primary sources of social science information, before linking up with international co-operative efforts aimed in particular at the coverage of secondary information sources.

Moreover, the question of information techniques and the compability of services is extremely important for small countries. These countries are above all interested in the development of internationally co-ordinated and compatible techniques and systems, as a means for partly or wholly overcoming their own traditional information self-sufficiency and isolation.

It is therefore very important for small countries to decide on the priorities to guide their international co-operation policy in the field of information.

The development of international co-operation depends on different objective factors, some of which have been mentioned in this article. But, if not more, an equally significant subjective factor concerns the attitude of information specialists participating in co-operative information efforts. It is imperative that information specialists acquire, early in their training, an attitude of co-operation instead of the still prevailing "do-it-yourself" outlook. Explicit institutional policy directives should reinforce the co-operative attitude of those actively engaged in the international transfer of information. The international exchange of experts and expertise is the pre-condition of an effective international transfer of information in general, and social science information in particular.

NOTES

1. See also "information Resources", papers of the International Conference on Information and Documentation in Social Sciences, Moscow, June 1977 (published in English in *Information Processing and Management*, Vol. 14, No. 3/4, 1978), especially contributions by J. Meyria and M. Gapochka on the co-operation among European and socialist countries (MISON) respectively.

2. For further information consult the work of Professor Jean Meyriat and the results of research by the authors of the present article. See also Item 88 in this issue.

3. A present-day example: 20 per cent of the whole book stock, or more than 250,000 volumes, of the library of the Kiel Institut für Weltwirtschaft derives from the exchange of publications. See p. 7 of the library of the institute's *1977 Report*.

4. Such co-operation can be traced back to 1902, when the plan for an international bibliography of the social sciences was put forward, on the one hand, by the International Bibliography Institute in Brussels, and, on the other, by the Hungarian Gyula Mandello and Ervin Szabó.

5. Such interrelationship has repeatedly been examined by ad hoc teams set up by Unesco. See also footnote 2 on page 265.

6. The Socialist Republic of Viet Nam has recently joined the system.

7. Meyriat and Gapochka, op. cit.

8. The conceptional and technical design of the DARE system are described in Unesco's "Reports and Papers in the Social Sciences", Nos. 27 (1973) and 31 (1975); the long-term strategy infrastructure of Interconcept are outlined in Unesco documents SS/77/Conf.601/2, December 1976, Annex E. An ECSSID-Interconcept meeting on terminological co-operation is foreseen in May 1981, at Bielefeld (Federal Republic of Germany). (See *ECSSID Bulletin*, Vol. 2, No. 2, 1980, p. 12–13.)

9. See also Item 116 in this issue–Ed.

10. Ibid.

11. FID publishes the annual *Extension and Corrections of the UDC and the P-notes* (the 30–40 issues per year contain proposals for revision) to inform UDC users about various actual and proposed changes within UDC.

12. See Item 116 in this issue–Ed.

13. The practical and theoretical aspects of the information in the social sciences as well as a publication programme have been discussed during the first meeting of the FID/SD held within the framework of the 40th FID Conference, Edinburgh, 15–16 August 1980. See also footnote 1 on page 265.

14. *Newsletter for Social Science Libraries*, Helsinki School of Economics Library, 1977–

15. International Association for Social Science Information Service and Technology (Association Internationale pour Les Services et Techniques d'Information en Sciences Sociales), *Newsletter*, Vol. 1, No. 1, November 1976. In English and French. See also the IASSIST Conference (held in conjunction with the International Sociological Association Congress, Uppsala, Sweden, 1978), *Proceedings*, e.g. "International Interchange of Data", 16–17 August 1978.

16. See Item 119 in this issue–Ed.

17. See also Item 88 in this issue–Ed.

18. The three-volume SPINES thesaurus was published in 1976; regularly updated, it permits the indexing of various aspects of science and technology policies. See also *Unesco Bulletin for Libraries*, Vol. 31, No. 4, July-August 1977, Item 209.

19. See *Unesco Bulletin for Libraries*, Vol. XXXI, No. 4, July-August 1977, Item 208.

In: Unesco Journal of Information Science, Librarianship and Archives Administration, 1980.4. 234–239.p.

ON SOME CRITERIA OF INFORMATION TRANSFER

The European Association of Information Service (EUSIDIC) was founded in 1970 for the purposes of information development by computers. It is now in the service of online information with a scope wider than Europe.

It is not frequent that a conference would bring together many intellectually strong as well as well funded institutions, although the two types of strengths are not mutually exclusive. There can be a concordance as the Conference of EUSIDIC[1] in Bled has shown.

Online

The conference was organised by the Slovenian Information Centre, Ljubljana,[2] on an impressive scale. Those taking part were: giant concerns or databank owning agencies, academical institutions such as: Shell, I.C.I., Lockheed, Systems Development Corporation, Chemical Abstracts, Electricité de France, Institute Français du Petrole, Unilever, Unesco, International Atomic Energy Agency, International Scientific Information Agency (Moscow), the Dutch National Library, CNRS-INFORMASCIENCE, the Polish Academy's Information Centre, the Hungarian Academy of Sciences.

Among the COMECON countries the Soviet Union, Czechoslovakia, Poland and Hungary took part for the first time in the conference the importance of which was emphasized by EUSIDIC organisers. No doubt it showed the possibilities of opening up an East European market.

What was on offer? A certain product called scientific information which is marketed by top level salesmanship, or rather, which is being transmitted from the producer to the consumer with the help of computers, undersea cables, space satellites with man-machine interaction. All these need great technical and financial investment (databanks, information channels) as well as increased intellectual capacities in information to develop interfacing.

This explains the presence in EUSIDIC of organisations such as Shell and Lockheed, or bureaux dealing with software such as the Schweizerische Institut für Technische Information or the Dutch National Library. The latter's experience is particularly interesting as its reference service has been online for years where experts assist the clients for a small fee. In other words it has an information/reference service which is based on a library.

Informatics and marketing

Above all EUSIDIC is a forum for online marketers and users which makes it a colossal scientific marketing organisation. This twaining was a new phenomenon in 1978 when EUSIDIC's long term policies were determined.[3]

The organisation deals directly with the development of information transfer in some and indirectly in other respects. Its main aim is to bridge the gap in information science between the US and Europe partly by linking up with US databanks and partly by intensifying European information transfer. According to the statement of perspectives, information is one of the world's resources which should be positively exploited internationally. In this connection one should recall the view by developing countries in 1979 at UNISIST II. The *information is the common inheritance* of the world. This is rarely acknowledged in words and never in deeds. The programme of EUSIDIC is realistic in aiming at a feasible budget by programming the services to fit the financial supplies.

According to the general programme the development of databases in the so called softdisciplines is less than adequate in Europe. See paragraph *B3* on databases which states that their access should not be restricted by geographical considerations and *B5* which states that the databases should not contain such personal or ideologically orientated material which was not intended for common use. The programme contains statements and proposes action relating to the access and uses of telecommunications networks.

In 1980 EUSIREF was formed by EUSIDIC with the cooperation of 14 countries, which is a new concentration of their information resources. There is a *EUSIDIC Database Guide*[4] with a monthly bulletin called *Newsidic*, apart from several other occasional publications which are available to members on a subsidy.

The Conference at Bled

On the 10th anniversary of EUSIDIC this conference staged twenty papers with question and answer sessions. There was no debate as such as some of the papers were preorientated and all of them were factually informative. The papers have been subsequently published in the 1980 plus issues of the *Online Review.*

Summarily speaking most papers indicated the general hope that online was to be cheaper, more accessible and less problematic technically speaking – with European postal services cooperating. Databases, information agencies and international organs keep offering their services to further that aim.

The present author has given an account that described three factors: his library in the context of providing primary and secondary information of multiple aspects from Mss to Scientometrics; he has also described the databases operating in his country and lastly, his country's participating potential in international information exchange.[5]

An open exchange of information?

In order to judge what relationship might exist between a nation's informatics and the international field the following should be noted:

In respect of scientific interchange our country is ranking about the twentieth while in its economic relations it might be about twentyfifth. In the European context it is rather a small country. Added to these one must add the isolation of its native language. The process we witness might be sketched thus:
linguistic isolation → open economic system → open science → open system for information.

From the above certain criteria can be deducted vis à vis the international exchange of information:

a) we have an interest both in the traditional and the computer based exchange of information;
b) home produced databanks are rarely needed in the humanities;
c) in science and technology we needed to use international databanks off or online;
d) off and online are not mutually exclusive but supportive;
e) whenever possible our national specific input should be offered; this has both scientific and economic advantages[6];
f) no databanks or viewdata systems should ever forget that the basis of scientific information is the library and its services;
g) the training of specialists in informatics is a key problem.

NOTES

1. *Newsidic* ISSN 0141–6243. It is actively supported by IIASA (International Institute for Applied Systems Analysis), Laxemburg, Austria.

2. EUSIDIC in association with the Information Centre, Ljubljana. Programme of the 1980 Anniversary Conference. Problems and Prospects of International Information Exchange. 23–26 September, 1980. Bled, Yugoslavia. 10 p.

3. EUSIDIC Policy Statement. (P.O. Box 85566, The Hague, Netherlands.)

4. EUSIDIC Database Guide. (1st ed.) Alex Tomberg. Oxford. 1979. Since then updated.

5. Some criteria for the participation in the international exchange of information: a Hungarian view. 8 p.

6. In the case of Chemical Abstracts the input of various countries like Britain, Japan are provided by the experts of those countries. Their work is reciprocated by services.

In: TMT, 1981.3.93–96.p.

EUROPEAN CO-OPERATION IN SOCIAL SCIENCE INFORMATION AND DOCUMENTATION: A PROCESS OF MATURATION

Social science information for all: some characteristics

Social science information is not aimed at social scientists alone, it is for everybody; it tends to become universal in one form or another through the media, much in the same way as the most advanced physics manifests itself in the form of electronic gadgets. Both, the repacked information, transmitted by the media, and the products of micro-electronics equally shape our views and way of life. Its global nature can be seen as another feature: the world's major problems, such as the upholding of peace, nutrition, energy, the struggle against terrorism, the supply of raw materials and the like, affect every country, and they are all of a social scientific character as regards their solutions. Related with the latter there is also a large-scale innovational process, necessary to the solution of global problems, which includes science, production and information, the latter being at once of a natural, technological and social science character. Though practically none of the so-called global problems can be solved without some contribution by the social sciences, it is generally valid, that research on universal, global and innovation problems is multidisciplinary in character, and so the corresponding information must be too.

Antecedents and aims of ECSSID

European Co-operation in Social Science Information and Documentation (ECSSID)[1] as an initiative institutionally to establish Europe-wide co-operation in social science information is undoubtedly linked with the Helsinki Final Act. The generator of the conception and its institutional base in the European Co-ordination Centre for Research and Documentation in Social Sciences, commonly known as the Vienna Centre, a non-governmental organization and an autonomous body of the International Social Science Council founded in 1962, and established in Vienna in 1963 under an agreement between Unesco and the Austrian Government.[2] For ten years the centre was subsidized by Unesco. Since this launching period, the centre has been supported by Unesco contracts, by the twenty-one member countries in various forms, and also by other bodies, such as Academies of Sciences, Unesco national commissions and the like. The main objective of the centre has been, and remains, the development of comparative research work.

For many years the documentation aspect hardly existed within the activities of the centre, it was no more than a name. But, to promote linkages between Eastern,

Western European and North American scientists through structured information channels, and to facilitate communication among the European social science information institutions the Board of Directors of the centre, at its thirteenth session in Paris, 1976, decided on the launching the ECSSID Project.[3]

One main difference between the ECSSID and similar, regional projects and organizations lies in the complexity of this project, the objective of which is rather ambitious. Its aims include the exchange of publications, information on on-going research, various aspects of automation (the problems of thesauri, compatibility of magnetic tapes, etc.), education and training, publishing activities, the organization of scientific conferences and workshops, various forms of exchange of experiences, and contributions to the development of national social science information and documentation (SSID) through bilateral and multilateral programmes, all in co-operation with existing similar programmes and organizations.[4]

This means that ECSSID in no way contradicts other international programmes, nor does it overlap their objectives.[5]

The structure, organization and activities of ECSSID

The current policy and management of ECSSID are carried out by the International Organizing Committee (IOC), whose members were initially nominated by the Board of Directors of the Vienna Centre, selected on a reasonable geographic basis, and then completed by the representatives of the national SSID services, focal-points of ECSSID, during the meeting of its major decision-making body, the General Conference.

The General Conference nominates working groups to discharge various activities, and publication boards to deal with the editorial work.

The structure of ECSSID is very flexible. In the centrally planned economies, it is usually the social science information centres of the Academies of Sciences that assume the role of "focal points" as main contributors to ECSSID activities. They play the same role within ECSSID as do the National Commissions within Unesco, or the UNISIST and the inclusion of social science information in such a programme has repeatedly been examined by ad hoc teams set up by Unesco. The focal points, on the other hand, are linked with the national committes of the Vienna Centre, and partly rely on their own budgets.

ECSSID meetings as a rule are financed by the host countries (focal points) as far as accommodation and organizational costs, and fees are concerned, while travel costs are covered by the participants.

Focal points are more familiar to the socialist countries with their Academies of Sciences than to countries without centrally planned economies and science, yet difficulties have always been solved despite the fact that in some countries even the role of focal points as national organizing bodies of SSID sometimes raise problems.

Focal points take the main responsibility of organizing, at the national level, the co-operation in social science information, thus contributing to ECSSID, and also have the task of contributing in various forms to the financing and assisting of activities related with the working groups, publications, meetings, etc.

The following ECSSID national focal points exist at present:[6]

Austria	Sozialwissenschaftliche Dokumentation der Kammer für Arbeiter und Angestellte für Wien, Vienna.
Bulgaria	Scientific Information Centre for Natural, Mathematical and Social Sciences, Bulgarian Academy of Sciences, Sofia.
Canada	Social Science Federation of Canada, Ottawa (Institutional partner of the Vienna Centre).
Czechoslovakia	Main Library, Centre of Scientific Information, Czechoslovak Academy of Sciences, Prague.
	Central Library of the Slovakian Academy of Sciences, Bratislava.
Denmark	The Royal Library. Copenhagen (contact point).
Federal Republic of Germany	Informationszentrum Sozialwissenschaften (IZ), Bonn.
France	Le Groupe des Sciences Politiques, Économiques et Sociales (SPES), Paris.
	National focal point including: Centre National de la Recherche Scientifique/Centre de Documentation des Sciences Humaines; Direction de la Documentation Française, Banque d'Information Politique et d'Actualité (BIPA); Fondation Nationale des Sciences Politiques, Services de Documentation; Institut National de la Statistique et des Études Économiques (INSEE), Départment de la Diffusion.
German Democratic Republic	Wissenschaftlicher Rat für Gesellschaftswissenschaftliche Information und Dokumentation bei der Akademie der Wissenschaften der DDR, Berlin.
Hungary	Library of the Hungarian Academy of Sciences, Budapest.
The Netherlands	Social-Wetenschappelijk Informatie en Documentatie Centrum (SWIDOC), Amsterdam. Koninklijke Bibliotheek, The Hague (at this stage a financing institution).
Norway	University of Bergen. Social Science Data Services.
Poland	Department of the Scientific Information of the Polish Academy of Sciences, Warsaw.

Romania	Office for Information and Documentation in Social and Political Sciences, Academy of Social and Political Sciences, Bucharest.
Spain	Departamento de Libro y Bibliotecas, Ministerio de Cultura, Madrid (contact point and a financing institution).
Sweden	The Research Council for Humanistic and Social Sciences, Stockholm (national focus agency);
USSR	The Institute of Scientific Information in Social Sciences (INION), Academy of Sciences of the USSR, Moscow

One crucial point, in addition to general international co-operation, within which ECSSID acts simultaneously as tool and product, is the extent to which the national focal points can be built up and are becoming suitable to encourage and organize co--operation within countries.

Achievements and problems

ECSSID relies on existing international co-operation programmes and does not overlap with them.

Together with its other characteristics this represents a tendency rather than a precise programme of action and applies primarily to the overlapping, parallel work. After all, one crucial feature of almost every programme of co-operation in information includes various forms of exchange of experiences, the promotion of information exchange, professional education and training, the introduction of up-to-date technology, and the creation of the intellectual bases for the latter: automation, compatibility in terminology and systematization, etc.

The specific character of ECSSID is not to explore *terrae incognitae* (still less to create them) but rather to apply information transfer on a regional basis to all social science, on the widest scale, from the promotion of exchange of primary documents to the building up of computerized data bases. At the same time, it is certainly designed to explore lacunae or tasks demanding greater efforts such as, exchange of information on ongoing research projects in social science.

The first "founding" conference, ECSSID 1, was convened by the Vienna Centre under the auspices of Unesco, prepared by the IOC meeting in Paris, and organized by INION (Institute of Social Science Information of the Academy of Sciences of the USSR) in Moscow, in June 1977, attended by specialists from nineteen European countries and Canada, and also by representatives of six international organizations.[7] Its main purpose was to form an overall picture of the situation of SSID in Europe, as well as of needs, potential and existing possibilities of co-operation. This survey resulted in a unique description of the situation, which has been published.[8] The recommendations of the conference include the establishment of working groups related to specific information topics (see Fig. 1) and the preparation of various documentation materials (e.g. bibliographies).

Unesco
linked mainly through its General Information Programme (PGI), and
its Sector of Social Sciences

International Social Science Council

Vienna Centre

ECSSID Project

General Conference
usually at two-year intervals

International Organizing Committee (IOC)
(acting through the technical assistance of the Vienna Centre's Secretariat
and providing for the continuing activities between General Conferences)

Working groups (WG)

1. Exchange of primary and secondary documents
2. Exchange of information on ongoing research
3. The compatibility of automated systems
4. Exchange of know-how; education and training of SSID specialists and users

Editorial Board

ECSSID Bulletin

(published in 4 issues a year by the Library of the Hungarian Academy of Sciences)

Social Sciences in Socialist Countries, Bibliography (comp. INION)

Ad hoc publications

(e.g. the so-called Euroguide on European SSID services, prepared by the Information Centre of the Bulgarian Academy of Sciences)

Occasional papers (not yet published)

Project of series under the control of the IOC, 2–3 items of about fifty to sixty pages per year on selected topics, (to be published by the Library of the Hungarian Academy of Sciences)

The structure and organization of ECSSID

ECSSID 2, supported by Unesco, was held at Blazejevko, Poznan (Poland) in October 1978, organized by the Polish focal point, the Department of Scientific Information of the Polish Academy of Sciences. Specialists from twenty-two countries and seven international organizations were present.[9] Its recommendations gave a precise definition of the objectives and range of activities of the project, as well as of the means of achieving the objectives; it also defined the joint responsibilities of the Vienna Centre and the national focal points (for example the Vienna Centre is responsible for the co-ordination of IOC activities, working groups and joint research projects of the programme, while the national bodies are responsible for the granting of fellowships or exchange of specialists); it established certain rules of procedure for the IOC; recommended that the national focal points provide the maximum material and financial support possible; stressed the necessity of having national focal points in each participating country. The conference also adopted the programme of activities for 1979/80.

ECSSID 3 was held in Bad Hennef (FRG) in 1981, while the 4th ECSSID was in Athens in 1984.

Working groups and publications

The bulk of ECSSID activities are carried out within the working groups and find expression in publications. East-West co-operation applies to the topics, to the experts, and also to the locations of meetings. Let me note that, in my opinion, two groups of activities are proceeding within ECSSID which can be qualified as specific both as regards their content and regional character: education and training in social science informatics, and tasks related with the building up of computerized social science data bases, or rather the application of informatics in social sciences on the widest scale possible. These are the very fields in which ECSSID can now produce the most novelties, but in this case, too, it relies heavily on existing programmes, national and international.

Working Group 1 is engaged in the regional extension of the exchange of primary and secondary documents. While this is the most traditional form of co-operation, it also has the greatest potentials which are to be explored and activated.

Working Group 2 deals with the exchange of information on ongoing research. Although Unesco and the Smithsonian Institution arranged a large-scale conference in the last quarter of 1975 in Paris on the worldwide exchange of information in this field, this being an important issue on the agenda of various European meetings, and organizational efforts have been made to promote regional co-operation, the results so far are rather meagre. The related programme, coordinated by the Dutch focal point, SWIDOC, is a step forward, although the planned first report – due perhaps to the delicate nature of the topic – will hardly cover all the European countries.[10]

The scope of Working Group 3 ranges from the intellectual tools (e.g. thesauri) needed for computerized data bases to the exchange of magnetic tapes. Various actions are proceeding in parallel and results can hardly be expected in the immediate future. Achieving compatibility of linguistic tools, terminologies, and different systems is extremely energy- and time-consuming work. Great efforts are being made to compare and make as compatible as possible the UDC (Universal Decimal Classification: Class 3, social sciences), the "Rubricator" of MISON, and the BSO (Broad System of Ordering) prepared by the International Federation for Documentation (FID). Although supported partly financially, partly professionally by several international organizations (FID, MISON, INFOTERM, International Committee for Social Science Information and Documentation), this work is hindered by many ideological, professionally conservative, traditional and linguistic obstacles. Unesco's pilot project INTERCONCEPT is another constituent element in this context.[11]

Working Group 4 was formed to develop co-operation in education and training of specialists in social science information, which seems to be one of the most promising ventures. It deals with the comparison of curricula with the preparation of recommendations for curricula, and especially with practical forms of training. Thus a week-long graduate seminar on social science data banks[12] was organized by the French focal point SPES and supported jointly by the French authorities and Unesco for some twenty-five participants. Further efforts along these lines will largely depend on the outcome of this novel undertaking.

Among the publications are those of current or permanent type such as series and periodicals, and ad hoc publications. They are all supervised by an international editorial board.

Among the ad hoc publications is *Euroguide*, an undertaking of the Scientific Information Centre of the Bulgarian Academy of Sciences to give comprehensive data on social science information institutions all over Europe.

Brought out as a follow-up of the recommendations of ECSSID 2 by the Library of the Hungarian Academy of Sciences, the *ECSSID Bulletin* appears in 4 issues a year, containing information on the development or structural, organizational and other changes in European SSID services, developments at national, international and regional levels and reports on ECSSID activities, as well as on the co-ordination work of the Vienna Centre.[13]

NOTES

1. A good part of Vol. II, No. 4, 1980, of the *Unesco Journal of Information Science, Librarianship and Archives Administration* was devoted to the problems of social science information and documentation. The issue contains a number of references to ECSSID in the following articles: Jean Meyriat, "International and Regional Co-operation in Social Science Documentation", pp. 227–233; György Rózsa and Tamás Földi, "International Co-operation and Trends in Social Science Information Transfer", pp. 234–239; Stephen C. Mills, "Regional Co-ordination in Social Science Documentation: the Vienna Centre", pp. 240–244. The following periodicals also include regular information on the ECSSID activities; *Vienna Centre Newsletter* (Vol. 1, No. 1, 1977) is published three times a year by the Vienna Centre, P.O.B. 974, 1001 Vienna, *ECSSID Bulletin* (Vol. 1, No. 1, 1979), published irregularly in practice four times a year, is devoted entirely to the ECSSID Project and published by the Library of the Hungarian Academy of Sciences with the professional assistance of the Vienna Centre, P.O.B. 7, H–1361 Budapest.

2. For current detailed information on the centre's activities with special respect to the ECSSID Projects, see *Vienna Centre Newsletter*.

3. The idea and experiment in international co-operation among social science information institutions has existed over the past thirty years, and several bodies have been involved in it. In this respect, special mention should be made of the International Committee for Social Science Information and Documentation, which has played a pioneer role. Cf. Rózsa and Földi, op. cit.

4. Cf. Meyriat, and Rózsa and Földi, op. cit.

5. To mention just a few FID (International Federation for Documentation) is a world wide and general organization for documentation with only a certain interest in SSID; IFLA (International Federation of Library Associations) similar to FID; IFDO (International Federation of Data Organizations for the Social Sciences); MISON (the SSDI organization of the Academies of Sciences of the socialist countries).

6. For addresses see: *ECSSID Bulletin*, Vol. 2, No. 1, 1980, pp. 26–27.

7. Austria, Belgium, Bulgaria, Canada, Czechoslovakia, Denmark, Finland, Federal Republic of Germany, German Democratic Republic, Hungary, Netherlands, Norway, Poland, Romania, Spain, Sweden, Switzerland, United Kingdom, USSR, Yugoslavia; also participating were Unesco, the Vienna Centre, United Nations Library at Geneva, International Committee for Social Science Information and Documentation (ICSSID), International Federation of Documentation (FID), and International Social Information System (MISON) of the socialist countries.

8. The conference papers were initially published in Russian and English versions by INION, and later as a special issue of *Information Processing and Management*, Vol. 14, No. 3/4, 1978, Oxford, Pergamon Press.

9. Austria, Belgium, Bulgaria, Canada, Czechoslovakia, Denmark, Finland, France, Federal Republic of Germany, German Democratic Republic, Greece, Hungary, Italy, Netherlands, Norway, Poland, Romania, Spain, Switzerland, United Kingdom, Yugoslavia; and Unesco, FID ICSSID, IFDO, IFLA, INFOTERM and MISON.

10. *Social Integration of Ethnic Minorities Including Migrant Workers*, Bonn, IZ. (Until 1987 5 vols. were published by the WG 2.)

11. Imre Molnár, *Establishing the model of INTERCONCEPT, International Terminological Information Network (INTERMIN) in Social Sciences, General Survey and Project*. Project manager: György Rózsa, Budapest, 1980, p. 112 (Unesco Contract No. 3671. IDS/32 270314).

12. The programme included: (a) a survey of social science data bases available in Europe or in some parts of Europe whether from European or overseas origin; (b) current policies of European countries referring to data bases in general, and specifically to social science data bases: their creation, financing, exploitation...; (c) an analysis of existing social science data bases: structure, coverage, service provided, condition of on-line and off-line access (from a technical as well as from an economic point of view), indexing and retrieval tools; (d) multidisciplinary data bases outside social science: their use for the social science; (e) specific characters of data bases in the social science and humanities: the role of national factors: political, legal, ideological, economic...; (f) linguistic barriers; progress being made towards common indexing languages and command languages; (g) networking: status and problems of West European networks; (h) networks in the USSR and the CMEA countries;

(i) regional and transnational exchange of data; problems of international co-operation between bases and networks; (j) concluding round table; the human factor; new functions of the information specialist, his training and status.

13. The *ECSSID Bulletin*, is free of charge and is printed in about 700 copies. It can be obtained on request from the Library of Hungarian Academy of Sciences P.O.B. 7, H–1361, Budapest. News and announcements by participating countries are welcome.

Bibliography of ECSSID publications 1977–1987:

1. *Papers of the International Conference on Information and Documentation*. INION, Moscow, 1977. Vol. 1, 301 pp., Vol. 2, 138 pp.
Reprinted in: *Information Processing and Management*. Vol. 14, No. 3/4, Pergamon Press, Oxford, 1978, pp. 138–318.

2. *Application of Mathematical Methods and Computers in Social Sciences. International Bibliographical Index*, 1975–1977. INION, Moscow, 1978, 268 pp.

3. *Social Sciences in Socialist Countries, Vol. 1, Selected Literature 1978–1979.* Ed. by INION under the direction of R. Mdivani, INION, Moscow, 1981, 265 pp.

4. *European Guide to Social Science Information and Documentation Services* (EUROGUIDE). S. Gabrovska, M. Biskup, A. Bossilkova (eds.). Pergamon Press, Oxford, 1982, 234 pp.

5. *Social Integration of Migrant Workers and other Ethnic Minorities. A Documentation of Current Research.* M. Herfurth and H. Hogeweg-de Haart (eds.). Pergamon Press, Oxford, 1982, 265 pp.

6. *Impact of Technology on Society. A Documentation of Current Research.* B. Schmeikal, H. Hogeweg-de Haart and W. Richter (eds.). Pergamon Press, Oxford, 1983, 259 pp.

7. *National Reports Presented to the IV ECSSID General Conference, Athens, Greece.* Ed. by K. Thesen Saelen. Bergen, Norway, 1984, 173 pp.

8. *The Changing Role of Women in Society. A Documentation of Current Research.* W. Richter, H. Hogeweg-de Haart, L. Kiuzadjan (eds.). Akademie-Verlag, Berlin, 1985, 819 pp.

9. *Social Sciences in Socialist Countries. Vol. 2. Selected Literature from 1980–1985.* Ed. by R. Mdivani, INION, Moscow, 1986, 343 pp.

10. *Impact of Technology on Society. A Documentation of Current Research.* A. Marks, B. Schmeikal-Frey, H. Hogeweg-de Haart (eds.) North-Holland Publishing Co., Amsterdam–Oxford–New York, 1987, 272 pp.

11. *Peace Research. A Documentation of Current Research (1983–1985).* L. Kiuzadjan, H. Hogeweg-de Haart, W. Richter (eds.) INION, Moscow, 1987, 316 pp.

In: International Social Science Journal, 1981.3. 559–565.p.

HOW TO BE AN INTERNATIONAL LIBRARIAN?

– Remembering the U.N. Library at Geneva, 1969–1976. –

Directing the U.N. Library at Geneva (formerly of the League of Nations) has, up till now, been the top library information function any Hungarian librarian has ever filled. The duration of six years has also been the longest. Geneva is central in the European consciousness, experience in its international library is noteworthy even for that reason only.[1] The combination of work and institution has made working there particularly interesting, even beyond personal recollections which authenticate the picture I drew seven years earlier.

Personal motives

The first title of the present paper was: *My good years in Geneva*... Yet, you may ask, if the years had been that good why did I not stay there and write an account as Chief Librarian at the U.N. and not as that of the Hungarian Academy Library?

When I left Hungary for Switzerland in 1969 my then superiors the President and the Secretary General of the Academy released me on unpaid leave with the promise that I had then been working for 9 years as the Academy's librarian having worked previously in the Library-Doc. of the Institute of Economics of the Academy. The formative experience of my working life was my work for the Academy and none of us thought that I should end up elsewhere. I fullfilled my obligation when out of two good positions I returned to the original one. I set the timing for my return to coincide with the 150th anniversary of the Academy; I left Hungary on 15 January 1969 and returned there on 15th September 1975.

Well, how does a Hungarian librarian become a Genovese internationalist? There are many factors and some go beyond personal qualities and qualifications. These are: regional and national provenance (there are geographical frameworks and other limitations; their prehistory of institutional position, where the vacancy occurs, e.g. how did the previous encumbent cope with his tasks and whether he was a "good diplomat" in relation to various delegations). The choice should be befitting to East-West relations and their balance of influences, and personal traits, including library experience, knowledge of language and some managerial and diplomatic skill.

Originally I was designated in 1967 to UNESCO to head their then new Department: Documentation-Bibliothèques-Archives (DBA).[2] In 1968 the offer for Geneva came from U.N. New York, via the Hungarian Ministry of Foreign Affairs. I went for an interview in Geneva and shortly afterwards was appointed Chief Librarian.

I could recall (for more background material) my original start as a diplomatist after the war, which ended with the Rajk trials, but really, much more readily, my contributions to the social sciences and to librarianship are the relevant ones. The milestones were these: research experience and publications, member- and Chairmanship of the International Committee for Social Science Documentation, information management with special emphasis on social sciences, my diploma in librarianship, my doctorate in economics and my work with national and international professional organs.

These are not prerequisites towards becoming an international librarian but possible avenues leading up to it. Since the present approach is an essay and not a history, I would like to recall what the U.N. has meant to me.

The U.N. from the cradle to the grave: shadows and illusions

From the time of its very foundation failure was dogging the U.N's heels escorted by two companions: illusion and realism. The League of Nations had been terminated by the war – that is the shadow; the illusion is that the world has learnt a lesson and the U.N. will be able to make things better.

The two sides of the coin

Nearly half a century of U.N. – this means something. The League of Nations had lasted only one score of years and its membership was waning all the time. The U.N. still keeps growing – not of its own volition – and expresses the tendency of the strengthening of international links. The changes mark the end of colonial empires, the birth of developing countries and the growing strength of the socialist block. The U.N. is one of the main channels and catalysts of international cooperation today.

The Secretariat is the U.N.'s executive organ, but the U.N. is what its members wish it to be, its policies are in the hands of governments. The executive organ can hardly make significant modifications of policy. The U.N. fails or succeeds through its members, its criticism is self-criticism. The more concrete the more important is the work of its branches, e.g. prevention of international drug-trafficking, promotion of the welfare of refugees, regulation of East-West trade, detente, etc.

Two points, inter alia: those who criticise the U.N. for its verbosity forget that all bureaucracies, national or international, operate with excessive documentation. Later on I shall discuss the second point which I call: anti-Coolidge effect.

The illusion that had enveloped the U.N. at its early stages gave way to a loss of illusions. Neither was justified. Perhaps the U.N. is like oxigen: its presence is self-evident, its loss signals danger. The world would be a more dangerous place without it.

Starting points and staying stages

During an international meeting I visited the Library once in 1963. Then and there I met Breycha-Vauthier, a well known personality of the library world, who headed the U.N. Library then. Later we met several times: in Budapest, in Vienna, and in Geneva. He was a warm and open man, who used to work for the League of Nations, opposed the Nazis, and was a member of the Knights of Malta. He first informed me that: "documentation is library work, badly done." Whether he had meant it or not, the position of the U.N. Library in 1969 was oozing of that mentality. Conservation prevailed. Breycha-Vauthier had already retired from his post in Geneva and worked as an Austrian diplomat in the Near East. The U.N. library staff had a reverence for the hierarchy. When, at an occasion of his Geneva visit I received him immediately, he admonished me. Who is so approachable – my visitor was saying – risks that others might think that he has little to do.

It was a mistake to leave the Library without a head for 5 years, but at length V. Winspeare-Guicciardi, the Deputy Secretary General of the U.N. decided to act. The Library needed a director because the morale was low, and professionally it was in a critical condition. The two depend on one another: there were personnel and personality problems I need not detail, but there were problems which I can summarise. Human relations, conceptual, organisational questions, points of initiatives, budgetary and staff problems.

The atmosphere was not very good to start with: factions, inner clashes and the like. The Library's reputation was at stake; the interregnum had brought increased insecurity. The staff – recruited from 18 different nations – was dismissed. There was no long term perspective to work towards, partly because the Library lacked initiative so its users did not put their trust in it. The U.N. directorate itself seemed to put up with a day to day existence.

The difficulties were increased by a lack of organisation and the routine. Although there were individual job descriptions, there was no organisational rule and only intermittant guidance for the use of the collections. The Dag Hammarskjold Library (DHL) in New York took the lion's share of the library service budget. The U.N. Library in Geneva was the poor relation, and to some extent it still is. Despite efforts under my directorship the Geneva Library remained underdeveloped both financially and staffwise. The promises, in the main, remained promises.

There was no budgetary provision for instance for official missions although some 2000 dollars were available due to the generosity of the U.N. Deputy Secretary General. This proved advantageous to New York DHL as well. It could be left to Geneva to represent the U.N. in European conferences.

Promotions had been neglected. This ill has largely been cured: new people were selected for new posts – documentation, reference work – but my choices weren't always lucky ones. My superior – who was very helpful all the time – had given me a free hand in recruiting.

The Library of the conference city

Geneva is a conference city ever since the time of the League of Nations. Housed in the Palais des Nations the Library expresses this characteristic and, no doubt, it is partly responsible for Geneva becoming a conference city. It is more than just a library; it is a centre for studies of international links, and, beyond its duties relating to the U.N., to delegations and to conferences, it is an international focus for social scientists.[3] It has a special place in the Swiss and in the European order of libraries, being a base for the practical study of international relations. While in New York the DHL may rely on the strength of the New York Public Library and Columbia University Library, in Europe the researchers and libraries look to Geneva in respect of the study of international links and relations.

The Geneva Library has a dual function: it is an official library yet it also serves a community of scholars. Its collecting sphere includes history, and the whole range of the social sciences. It is an integrated body with archival, museological, information and library functions. This had struck me as not dissimilar to the Hungarian Academy Library, which also draws two types of users: those in official and those in some scientific capacity.

The Library was founded together with the League of Nations in 1919, and in 1920 it was transferred from London to Geneva. (This continuity with its mother institution is another similarity with the Hungarian Academy Library.) The Palais was built in 1936.

In 1927 the L of N received the fund of $ 2 million from John D. Rockefeller, so that it might become the centre for international links. Up to the present day the Library enjoys the interests of this sum.

A sketch plan for the activities of the library

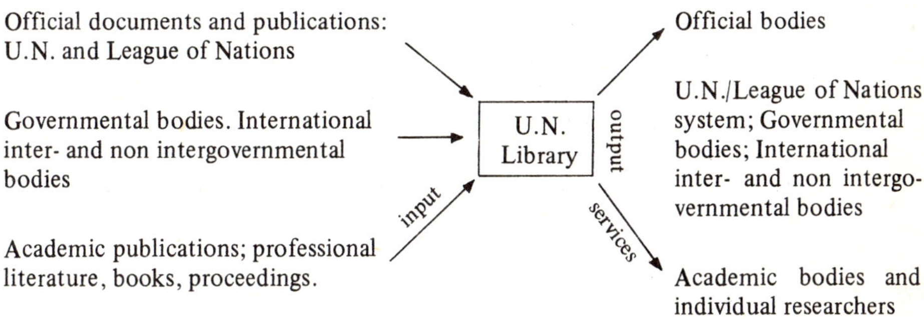

There had been 800,000 volumes of books, 10,000 current publications and several millions of other documents, filling, in toto, 35 km of shelving. There were thousands of books and periodicals in the subject reading rooms (international documents, periodic publications, social science collections) which, on the whole, gives an overview of representative collections.

From reference questions to in depth literature searches, the Library was obliged to take on everything. It served the 3000 officials of the Palais as well as the participants of the congresses. In 1974 alone there were 5600 conferences held in the Palais, but the Library also served "outside" conferences (e.g. the Security Conference) as well.

The two basic publications of the Library that started in the twenties, were: The monthly list of books catalogued and The monthly list of selected articles. These contained about 13,000 headings annually.

The catalogue of periodicals started in my time, in 1972. In 1970 we had started a bibliographical series called: Reference lists. Our bibliographies on East-West trade and on international law were received favourably. The U.N. and the various governments were appreciative of the microcard bibliography, edited in 1974, which listed the 8000 U.N. documents on the Near East.

Stepping out from its relative isolation the Library contributed to the dissemination of documents and to the know-how of documentation, in conjunction with UNITAR (U.N. Institute for Training and Research) and FID. In 1972 it assisted in the organisation of the First World Symposium of International Documentation in Geneva.[4]

The League of Nations archival collection was the basis of the historical collection which was described by the Guide to the Archives of the League of Nations and to the Archives de la Société des Nations Répertoire général 1919–1946, vols. 1–3.

The snapshot and the perspectives

The previous chapter described the functions and the problems of the Library. The following official document summarize work and aspirations in Geneva IC/Geneve/2096, 22. Sept. 1975.

There have been three documents. Their dates should also be noted: the first development plan of the Library (1969–1975) was ready on 12th August 1969. The date of the Summary of its realization and my proposals was submitted on 8th September 1975.* It contains the main points of the Development Plan. These two were internal documents. On 22 September 1975 the Director General countersigned the Rules for the use of the Library, which was the first of such in the Library's history. His friendly gesture was characteristic: I had signed the document on 12th September (two days before my departure) and he countersigned it with a preface, which made it official.

* G/AD321/1(1)DFR–CONFI

ST

UNITED NATIONS

SECRETARIAT

IC/Geneva/2096
22 September 1975

Distribution: 1 copy per member of the Secretariat and Permanent Delegations

CIRCULAR No. 2096

Subject: USE OF THE VARIOUS SERVICES OF THE PALAIS DES NATIONS LIBRARY

I wish to draw the attention of members of the Secretariat, of Permanent Delegations and of other users of the Library of the United Nations Office at Geneva, to the rules concerning the use of its various services. The instructions contained in this circular cancel and replace those which have appeared in earlier circulars (IC/Geneva/871, dated 9 July 1962).

V. WINSPEARE GUICCIARDI
Director General

GE. 75–9683

12 September 1975
IC/Geneva/2096

To: Members of the Secretariat and Permanent Delegations

Subject: Use of the various services of the Palais des Nations Library

RULES AND INFORMATION CONCERNING THE USE OF THE LIBRARY OF THE UNITED NATIONS OFFICE AT GENEVA

1. The purpose of this document is to describe the services of the Palais des Nations Library, and to define the conditions for their use.

2. Authorized users

(a) *Official users:* staff members of the Secretariat, members of Permanent Missions, the staff of specialized agencies of the United Nations at Geneva, journalists accredited to the Organization, delegates of Conferences and consultants to the Secretariat.

(b) *Outside readers:* the Library, as an international research centre, is open to persons engaged in research, on presentation of a reader's card which is issued by the Library itself. The Library may require applicants for a reader's card to submit a recommendation in support of their application.

(c) *Users of the League of Nations Archives:* access is governed by the rules issued by the Secretary-General of the United Nations (see Circular ST/SGB/135/Corr.1 of 26 December 1969).

3. Consultation of the collections

Official users may either consult works in the reading rooms of the Library itself, or borrow works from the Library; outside readers may consult works only in the Library reading rooms. The following works may not be borrowed: works belonging to the collections in the reading rooms, United Nations documents, old, rare or valuable books and other publications requiring special protection. However, the Chief Librarian or officers designated by him may permit exceptions to this rule.

The reading rooms are open countinuously from 3.30 a.m. to 5.30 p.m., and works may borrowed at any time between 9.30 a.m. and 12 p.m., and between 2 p.m. and 4.30 p.m.

Access to the stacks is prohibited.

4. Loans

(a) The Library lends its works *directly* to users in the following categories:
– staff of the United Nations Secretariat holding regular appointments;
– members of Permanent Missions and accredited journalists.

IC/Geneva/2096
(page 2)

(b) The Library also lends its works, through *officers appointed for the purpose*, to users in the following categories:

– staff of the specialized agencies;
– delegates to Conferences;
– consultants to the Secretariat and temporary staff.

Reference units may transmit to the Library requests for loans, and the Library will lend the book in the name of the actual person who has requested it. Staff members of the Secretariat *are requested, preferably, to come to the Library in person*. This simplifies the search for works, which can therefore be obtained more quickly. However, in certain cases it will be possible for Divisions, by agreement with the Library, to operate the indirect loan system through their reference units. In these cases, the reference units or officers in charge of loans will assume entire material responsibility for the works loaned.

(c) The loan period for *books* is *one month* and, for *periodicals*, it is *15 days*. In both cases, loans are *renewable* before the date of expiry at the request of the borrower, provided that no other staff member has expressed a wish to consult the work in question. For technical reasons, the Library may recall any work before the expiry of the loan period. Borrowers must comply immediately with a recall notice from the Library.

5. Circulation of periodicals

The Library circulates about 3,000 periodicals through reference units. Although this system is to continue in force for the time being, the Library intends to *review* the circulation procedure in order to remedy certain anomalies which have arisen. At present, many staff members are not receiving the periodicals they have requested, and many issues of periodicals are not being returned to the Library. Also, some staff are therefore requested not to overtax the facilities of the Library, so that the services it provides may be distributed equally among all users.

6. Greater protection of the collections and safeguarding of readers' interests

The measures under consideration for improving the loan system and for reviewing the circulation procedure are designed to serve the interests both of the Organization and of users. The Library also has a duty to protect its collections, which constitute part of the Organization's property. Staff members are therefore reminded that, under the administrative rules, they are *materially responsible* for damage caused to property of the Organization (see Rule 112.3 of the Staff Rules, and the Organization's property). These provisions *apply to works borrowed from the Library*. The borrower of a book remains responsible for it until it is returned unless, with the agreement of the Library, it has been transferred to a third person. If a work is lost or is not returned after two recall notices have been issued, the reader is billed for the whole cost.

IC/Geneva/2096
(page 3)

Readers are requested to return to the Library, within the prescribed time-limits, all borrowed works which they no longer need. Users who are about the leave Geneva on long missions or on leave must return to the Library any works they have borrowed. Reference units are asked to assist the Library in recovering works left in offices.

7. Reference service

It may be useful to describe briefly how the Library's reference and information system operates.

(a) *A general reference unit* (room B.121, extension 4194) *provides information to users* on all questions of a general nature (conditions of consultation, location of the various collections, etc.), assists readers in using the main catalogue and in collecting reference works, and provides guidance to readers by telling them to whom they should address particular questions;

(b) *The loans service* (room B.121, extension 4184) deals with all questions relating to loans, their renewal, recall notices and inter-Library loans;

(c) *The Chief of the Readers' services and Documentation Section* (room B.46, extension 2291) helps users to compile specialized bibliographical lists. Many such lists have already been compiled.

(d) *The specialized reading rooms* contain collections of basic works, specialized reference works (bibliographies, encyclopaedias, etc.), current periodicals and *card indexes of articles in periodicals*. Librarians specialized in *law, economics, international organizations*, etc., are available to assist users in their research.

The reading rooms are as follows:

- *United Nations Collections* (room B.127, extensions 4185, 4186, 4187);
- *Economics, Finance and Transport* (room B.340, extension 4198);
- *Legal and Political* (room B.358, extension 4195);
- *Social Questions* (room B.352, extension 4196);
- *Specialized agencies* (room B.348, extension 4190);
- *Bibliographical Collections, General Current Press* (room B.135 /former Room XV/, extension 4015);
- *Historical Collections, League of Nations Archives* (room B.34, extension 3054, 3052).

(e) *The Acquisition unit* (room B.50, extension 2285) receives suggestions from readers for the purchase of books and provides information on current acquisitions. Those interested may here consult *publishers' catalogues*, brochures on new works, etc.

(f) *Publications*

Members of the Secretariat are reminded that the Library issues the following bibliographical publications:

IC/Geneva/2096
(page 4)

- Monthly List of Books Catalogued (bibliographical descriptions and analysis by subject).
- Monthly List of Selected Articles (a large collection of articles from about 1,200 periodicals).
- Reference lists (subject bibliographies, e.g., No. 1, "East-West Trade", No. 8, "International Co-operative Research", etc.).
- Daily Reference List of Newly Acquired Books, etc.; this may also be supplied in perforated form so that subject indexes can be formed.
- Catalogue of Periodicals... (volume containing analytical lists of periodicals, of series, of current yearbooks received by the Library, the total number of items being about 10,000).

8. Chiefs of divisions, the reference units under their supervision and all staff members are invited to co-operate to a greater extent with the Library in order to enable it to improve the services it provides to users.

György Rózsa
Chief Librarian

The professionalisation of political questions

One has to tread cautiously here. Anatole France said somewhere that his favourite reading was the list of books. Evidently, he has not worked in an international intergovernmental library, otherwise he would have known that even catalogues are not innocent tools. For example: Once a U.N. delegate complained on behalf of his government's foreign ministry that his country's professional literature was under-represented in the Library. I showed him the list of periodicals in our Monthly List which revealed that his country was adequately represented. In terms of books I requested that he should have a list of the most important ones published. This we compared with those in our catalogue. But before all this could catapult into blames and counterblames I explained that the nature of such complaints could have originated with certain authors who were, at the same time, delegates of conferences. When such authors visit a library they look their own names up in the catalogue. If they could not locate themselves in the catalogue they were ready to jump to conclusions.

The delegate then asked me how do I have such psychological insight. I told him that work in my Academy Library had equipped me dealing with such problems. He laughed and sent a favourable report to his government.

One of my own staff mentioned once that geographical denominations were out of date in our catalogue and certain countries could take offence. It was suggested that he should act. Truly, the catalogue was full of outdated information and as new countries were being born, their numbers increased annually. It would have been impossible to correct all catalogue cards. We have solved the problem by creating "see also" references after the new name of each country.

Once a delegate complained that we had exhibited the title page of a book that denigrated the head of his state. Well, he had a point. We acted on it, internally.

There were demands too, that the Library should get rid of certain books of political science, mainly memoirs. We had to decide on the information contents of such works, and if that was considerable, the book in question had a place in our collections. Despite of our caution, certain bibliographies of ours had non-desirable names in them. On the whole I succesfully resisted undue outside pressure, and the U.N. Office was never requested to intervene. A French expression might stand here as my Geneva motto: When I expressed my intention to leave I was requested to stay – perhaps because I managed to manage "sans histoire".

The anti-Coolidge effect

Documents are the bread and butter of the U.N. Library – indescribable quantity. The United Nations Document Index (UNDEX) and the United Nations Bibliographical Information System (UNBIS) both issued by DHL should have accounted for all the documents. With its manual index the Geneva Library wondrously managed to maintain an adequate finding tool. It coped with tens of thousands of working papers too, as was demonstrated by the quick production of an U.N. document index for the first Middle--East Meeting in Geneva (Gromiko–Kissinger).

The library's largest reading room housed – by way of relocation – the U.N. and its Specialized Agencies' collections.

The U.N. documents are verbose, hence my "anti-Coolodge" name for them. President Coolidge was famous for his short and precise expressions. Once on a Sunday, when he returned from church, his wife asked him what had happened there: "Preaching", he answered. "About what?", asked his wife curiously. Coolidge thought, then said, after a while: "Evil". Mrs. Coolidge was exasperated. "But what was the preacher's attitude about evil?" Coolidge pondered and said: "Disapproval".

International documents have anti-Coolidge characteristics. What can be said on a single page is said on ten, all full of gobbledegook. Having said that, the U.N. documents are indispensable and are the life and blood of the organisation.

Beyond the doors of the Library

The Library that had been known only for its collections was now beginning to make an impact professionally, too. It took over U.N. representation in European Conferences, published the Reference List, participated in the UNITAR Seminars and organized the 1972 Documentation Conference and issued a volume by the Document Processing Department.[5] The Documentation Conference featured such international experts as Pierre Piganiol, Franco Casadio, Peter Judge, John Goomarghtigh, Jacques Tocatlian, Sven Welander – representing the different branches of scholarship and of the information science.

Members of the Library took part in various international conferences, normally initiated by DHL. This was partly due to Lev Vladimirov, then director of the DHL, now head of library science at Vilnius University – a friend even today.

Our staff presented papers on the course of Societa Italiana per Relazioni Internazionali, to the meeting of the Association of Law Librarians and the International Council of Archives.

There were many others whose help proved immeasurable for me in my effort to fit in with a new environment but I should not start listing their names here.

Ici Genève...

A few words of the life of a city whose railway station the Cornavin should bear this inscription: "Ici Genève, tout le monde descend". I had grown fond of that town which feeling was never tainted with any dissonant experience.

It was not a bad thing to be a "V.I.P." in Geneva, but beyond that, private life was easy going, full of excursions to France, to the heart of Switzerland and twice yearly to Budapest. Nevertheless, I missed old friends, some of whom came to visit me in the Library.

Finally, what was gained by the Geneva experience, beyond the personal and the family advantages of travel and language learning? A multifaceted world provides chances to have a multiplicity of human contacts. Professionally it equips one to understand fully the world of official documents and their environment. It enables one to exercise professionalism and diplomacy together. It equips one to compare various types of daily routine work, sundry approaches and human attitudes. It enables one to polish one's committee technique. All of these can be gains if one wanted them to be.

For all its colour the daily work in Geneva was, perhaps, less interesting than my work in the Academy. On home ground one has a feeling of being involved in more important actions than in the U.N. where the final decisions are removed into the fora of multivarious committees. They decide on questions which they do not have the chance to examine in depth. The pyramidal structure gives a certain feeling of security but it discourages initiatives. (An organ does not need to be international for that.)

"How to be an international librarian?" By combining the national with the international. It is not an "ubi bene ibi patria" for me but it is – despite the red tape and the cynicism – an international attitude which is a trust in matters becoming better. The basis is the affection and the commitment one feels towards one's own country. I do not believe in the U.N. world citizenship but I do believe that the world is a better place for the U.N. It seems to me that the quality of the world today is, to a large extent, determined by the positive attitudes of its officials. I hoped to do my work in Geneva in this spirit and left the town and the U.N. with a good feeling.

NOTES

1. This paper was initially motivated by the 60th anniversary *Festschrift* planned for 1979. All former directors of the U.N. Library had been asked to send a paper describing their experiences. I sent a 60 pages manuscript for this purpose. Unfortunately, the *Festschrift* has not yet been published.

2. With the combination of the two major information programmes, UNISIST and NATIS into the *General Information Programme* (PGI – French abbreviation) the proposed integration was achieved just after two years.

3. Cf. Rózsa, G.: United Nations Library at Geneva: an international research centre. International Library Review. 1976. No. 2. 119–126. p.

4. Sources, organization, utilization of international documentation. Proceedings of the International Symposium... held in Geneva 21–23. August 1972. The Hague, 1974., 586 p. FID Publ. 506.

5. Documents of international organizations. A bibliographical handbook... Comp. and ed. by Th. D. Dimitrov. London–Chicago, 1973. American Library Association. 301 p. International University Publications.

INTERNATIONAL CONNECTIONS AND THE POSSIBILITY OF CO-OPERATIONS IN STOCKBUILDING

1. Some general points

Due to the expansion of professional literature comprehensive stockbuilding became illusory. This is true for countries as well as for institutions.

Some approaches:

1) Goethe's dictum whereby national literatures will add up to a world literature at the end of the 20th century relates to information in the sense that it has by origin and use a world-wide character.
2) If the information has a world-wide character then its source the primary literature needs an international approach.
3) The approach is based on multiple experiences of many years. There isn't one definitc way but there are several possible ways.
4) These can be categorised into two camps: centralised and decentralised stockbuilding. The British Library Document Supply Centre belongs to the first, the SCANDIA-plan to the second category.
5) Economical and political development indicate integration too, see the regional co-operations, the multinationals or the agencies under the U.N. Even information is integrative in tendency.
6) Within this tendency there are differences which crystallise both by scientific fields and by certain publications.

2. Stockbuilding and information

The world-wide character of information is evident in the physical and technical sciences and less evident in the social sciences and humanities. The first recognise no political or ideological boundaries, and such restrictions that there are, relate to the different types of culture in some countries. There is no sense, for instance, in building luxory cars for countries without hard surface roads or export pigs to countries where they don't eat their meat.

Among the humanities certain branches are national in character e.g. history of literature. Yet, in principle, the use of such literature has no limits, the information generated by it may be of international interest.

In order to represent national literature on an international level each country should build up a body of reference literature – on a shared basis. Within this framework, large catalogues, such as that of the Library of Congress should appear alongside bibliographical checklists of national or international character. National registers, lists of addresses, centralised catalogues are means to guide the reader to primary literature.

If mutual interests support it, the national development of information pools must precede their international sharing.

3. National needs for international cooperation

The national programme should include communication techniques, telephone, telex, electronic mail, etc. Another criteria is to agree on the modes of cooperation.

Effects of cooperation in acquisition which do not build on mutual interests are futile and seldom get further than declarations of interest. Acquisitions policies could only be pursued permanently with the well known budgetary and space constrains if the given library may rely upon an internal and international interlibrary loan system.

Cooperation on the field of information is easier than in stockbuilding. Cooperative acquisition is not even a national scale reality yet – mainly due to finances. There are, however, examples of super-national cooperation such as in Scandinavia where there is a regional scheme for sciences. Library groupings in a given country are not infrequent for cooperative acquisition or cataloguing (U.K., U.S.A.) aided by the computer. E.g. SWALCAP, BLCMP. In mid and Eastern Europe one may refer to MISON which is the cooperative venture of Scientific Academies in socialist countries for the promotion of international social science information. The union list of their social science serials has already been published as the starting point of a further range of publications. Depending on te development of communication and the exchange of information, stockbuilding may reach an international stage where the emphasis will be laid on scientific method rather than on administrative modes. It will span to books, periodicals and reports. The necessity to move cooperatively will, perhaps, be recognised in broader areas of science too. The need is already in existence and signs show that it will further increase. Conspectus in Canada, Britain and Holland show that resource identification is a means of coordinated collection developments.

The growth of needs will squeeze out cooperation. Let us not repeat the motto "one product – multiple use" automatically. The user is the significant factor and the continuity of large collections. In stockbuilding cooperation will necessarily be restricted mainly by traditions. Yet whatever is possible should be inbuilt and never carried out by campaigns or dictates. In 1983 there were administrative periodical reduction campaigns in various countries, Britain, Hungary amongst them. The Hungarian example should warn others that enforced cuts may be counterproductive.

4. Pertaining questions

Cooperation in acquisition cannot be limited to the coordination of buying and exchanging national and international documents. The principle of a network is part of the stockbuilding scene whether by way of voluntary cooperation as in the U.S. or by state agreements as in the socialist countries or through international cooperation as in Scandinavia. Cooperative acquisition is a reality in Scottish universities and West German libraries. But the principle goes well beyond the framework of coordinated stockbuilding.

The use of space is the first adjoining problem. One must consider stores and particularly outside stores. One must remember that to the purchase value of a book one must add its processing and shelving costs. In a large number of cases the costs of the latter exceed the former. Consequently the national and international coordination of stockbuilding aims not so much at reducing costs but at optimising the supply of information. Stockbuilding is a part of this effort, other parts being, reduction of certain divisions of stock, optimal utilisation of space, rationalisation of bibliographical control. Each would have its own experts literature.

In the work of establishing records one would increasingly rely on the inbuying, sharing or exchanging machine readable records, with obvious advantages. By now these are the norm.

International standardisation (e.g. cataloguing) should also be part of this scene with computer cataloguing and classification. Traditional classification will still survive and systems, like UDC, will keep on being revised. Processing procedures are under review too. The building and use of thesauri is also relevant in this context, not to mention conspectus development whereby libraries match their own profiles to one another.

5. Synoptic table

The illustration shows the possibilities of cooperation in co-related collection development.

In parting, one must emphasize the first criterion of coordinated stockbuilding which is the development of international relations in terms of economics, politics and culture. The European Culture Forum, Budapest, 1985, was a further step in this direction, the initial step being the Helsinki agreement.

Synoptic table

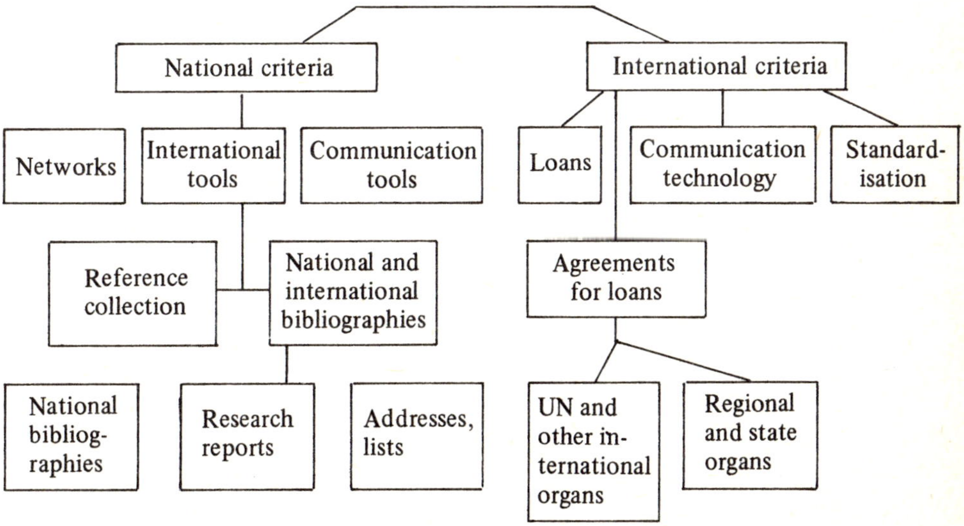

* In: Publicationes No.16. Library of the Hung. Acad. of Sciences. Budapest, 1987. 152–155.p.

100.000.000 BIT/sec
INFORMATION

III. "THE ELEPHANT'S HEAD" AND INTEGRATED INFRASTRUCTURE FOR DEVELOPING COUNTRIES

A CONCEPT OF AN EFFECTIVE MANAGEMENT OF INFORMATION TRANSFER TO DEVELOPING COUNTRIES WITH SPECIAL REGARD TO THE UNITED NATIONS

AN APPROACH TO AN INTEGRATED POLICY*

---

* This study is based on a "conference room" paper produced by the author, commissioned by the UN Office of Science and Technology for the Advisory Committee on the Application of Science and Technology to Development 22. session, Geneva, 15–16 Nov. 1976. – E/AC.52/XXII/CRP.1(Prov.).

Preamble

I. *Conceptual changes by the UN in information policy for development*

The formulation of the problem of information transfer. An integrated approach to information policy.

II. *The typology of the UN information systems*

The criteria and categorization of information functions.

III. *How is information managed by the UN?*

A simplified scheme of the information policy and its management and financing by the United Nations. Scheme A.

IV. *An attempt to integrate UN information systems into a real network*

Some steps to change the situation. The schemes of an integrated United Nations information system: B–F.

V. *A new approach to information transfer for developing countries: the Applied Information Centres – AIC*

AICs as "focal points". The prototype of an AIC. Scheme G. for the setting up of AICs.

VI. *Conclusions*

Notes

Preamble

The author is convinced that taking into account the results obtained so far in the matter of development information, high-level division of labour in information is not applicable to developing countries. New methods with more successful means of implementation would have to be found for the Third World.

Although the original paper for this study was written in 1976 the author considers that the main ideas of this paper are still valid since the successful implementation of an information system for the Third World is still a concept of the future. Basically, the suggestion for further development for the information supply of developing countries can be summed up in two words: *integrated policy*. Integration here has two meanings: *first*, integration as an UN policy combining the different developing information systems of the UN family; *second*, an integrated approach to the contents of developing information, which combines information supplies with education, manpower training, and with practical works (e.g. agricultural production, small industry, etc.).

I. Conceptual changes by the UN in information policy for development

Following the adoption of resolution No. 1902 (LVII) by the Economic and Social Council at its fifty-seventh session on 1st August 1974, concerning "the progressive establishment of an international information exchange system for the transfer and assessment of technology for the benefit of developing countries,"[1] a series of measures have been taken by the UN. Fundamental in the Economic and Social Council resolution is the assumption that the current UN scientific and technical information systems and services are not adequately meeting the requirements and demands of developing countries,[2] as stated in the study on "Improving technological information exchange with developing countries"[3] also prepared in response to the Economic and Social Council resolution 1902. Based on this study, a note was presented to the UN secretariat to the twenty-first session of ACAST.[4]

In its resolution A/RES/3362(S–VII)[5, 6] of the seventh special session, the UN General Assembly took a position on the question of development, international economic cooperation and the role of information. Its chapter III Science and Technology, para. 4, points out that "Developed countries should facilitate access of the developing countries on favourable terms and conditions, and on an urgent basis, to *informatique*, to relevant information on advanced and other technologies...", and in para. 4 of the same chapter that "The UN system should play a major role... to ensure the application of science and technology to development" and that "the Secretary-General of the UN should take steps to ensure that the technology and experience available within the UN system is widely disseminated and readily available to the developing countries in need of it."

Resolution A/3362 (S–VII) was preceded by a resolution on the new international economic order,[7] "The International Strategy for the Second UN Development

Decade",[8] and also took into account the report of the Group of Experts on reconstructing the economic and social sectors of the UN system.[9]

Based on the above-cited documents and related resolutions the UN convened the UN Conference on *Science and Technology for Development* (UNCSTD) 20–29 August 1979 in Vienna, the *ACAST colloquium "Science, Technology and Society: Needs, Changes and Limitations"* (14 working groups – the 11th WG: Information Systems in Science and Technology – with the participation of 450 specialists from 80 countries), Vienna, 13–17 August 1979[10] and the *NGO Forum "Science and Technology for Development"* 19–29 August 1979.

These three Vienna meetings have produced the most detailed documentation on science and technology for developing countries, including a previous important documentation on information.[11]

It should be mentioned that since the Vienna Conference there have been two newsletters which refer to the relevant documents, facts, resolutions concerning science and technology factors as well as information in developing countries.[12]

The above-cited UN documents and their intention should be considered as the main background for this study on information and development. It is aimed at analysing the status of the management of information transfer by the UN with the view to provide *better services for users* (governments, developmental institutions, international organisations) by suggesting some approaches to the integrated information policy and also to the searching of ways and methods in order to implement this policy. The justification of any information activity in and for developing countries lies in its potential to reduce technological and social dependence. One important factor in reducing this dependence is a more economical, more rational use of information. This furthers the better utilisation and exploitation of all resources affecting development. It also implies a concentration of efforts in the field of information as opposed to the existing proliferation of it. Concentration does not mean unification or an administrative centralisation: what it primarily means is a *conceptual change* in information policy and management, implying a real and harmonious integration of information activities for development. Within the framework of this integrated activity, a *real partnership* is to be developed between developing and industrialised countries. In this sense the UN may play the role of a bridge.

For this purpose, the study takes into account the existing international mechanism of the transfer of information, the major realized international co-operations in this field; first of all the UNESCO's PGI (Programme Général d'Information)[13], the SPINES-system launched by UNESCO, the DEVSIS and other programmes reflected by the *Indexing Vocabularies Produced by Organizations of the UN System.*[14] The present study concentrates on the problems of information transfer.

What holds true for the whole problem of development in general, also holds true for the particular field of information transfer. With respect to development, Gunnar Myrdal points out that what is needed: "a broader approach which takes into account the social reality of institutions and the attitudes formed within them, which attitudes, in turn, support them".[15]

The "broader approach which takes into account the social reality" in the field of information means the *reconsideration* of the existing international information assistance and systems for development (how to provide information) on the one hand, and a *search* for new ways and means, and methods for the infrastructure of information (how to assimilate, how to apply) in developing countries, on the other.

One may agree with Myrdal in his statement that "Development must be understood as the movement upward of the entire social system, where there is a circular causation between conditions and changes with cumulative effects." and "... Reforms must be directed upward moving the system upward as much and as rapidly as possible by inducing changes planned with this result in mind."

Information as an integral part of development activities should be considered as a facet of the "circular causation", and as a catalyzing factor of the "cumulative effects".

An integrated approach to the problems of information policy for development by the United Nations system can be one of the important steps toward the realization of "cumulative effects" to the benefit of developing countries.

II. The typology of the UN information systems

There may be different approaches to the typology of information systems. The first task, however, is to define what is to be meant by the term "information", and how this notion is interpreted in this study.

Criteria and categorization of the UN information functions

The word "information" used in the sense of scientific information has two meanings. First it means an *activity*. This is an organized process of transmission, or transfer of secondary information. Second, it denotes the *result* of this process, embodied in an intellectual product.

The Directory of United Nations data bases and information systems surveys 615 UN information systems.[16] In fact, the number of information programmes, systems and services depends on the starting-point of such a survey; thus their number may be higher or lower.

In the following brief survey, mention will be made only of the major international information systems, programmes and services of the United Nations family which are concerned with information for development.[17]

AGRIS/FAO:	International information system for the agricultural sciences and technology
CARIS/FAO:	Current Agricultural Research Information System – on on-going researches
CORE/IOB:	Common Register on Development Activities – on on-going projects
DARE/UNESCO:	Data retrieval system for documentation in the social and human sciences

DEVSIS/sponsored: by several organizations within or without UN	Development Sciences Information System
INDIS/UNIDO:	Industrial Information System
INIS/IAEA:	International Nuclear Information System
INPADOC/WIPO:	International Patent Documentation
IRS/UNEP:	International Referral Service for Environment
ISIS/ILO:	Integrated Set of Information Systems – a package of computer programmes (used by several organizations)
SPINES/UNESCO:	Science and Technology Policies Information Exchange System
UNISIST/UNESCO:	World Science Information System (it would be better to call it a Programme) – a normative programme sponsored also by the ICSU (International Council of Scientific Unions)
UNBIS/Dag Hammarskjöld Library:	United Nations Bibliographical Information System

The above-listed information organisms are representing a variety of types of systems/programmes/services. Most of them are already operational (AGRIS, INIS, DEVSIS, etc.), all of them are computer-based and are operating on a world-wide scale.

Apart from those mentioned above, there are several computer-based systems of *technical character*, not affecting the substantive problems of information, like the ISDS (International Serials Data System), the ISORID (International Information System on Research in Documentation) and services which are concerned with a region like the CLADES/ECLA (Centro Latino-Americano de Documentación Económica y Social) or are not directly related to the transfer of information for development. These services are highly useful for development information, too, like the Monthly List of Selected Articles produced conventionally by the UN Library in Geneva.

Information systems in general can be categorized by their major tasks, and by their main characteristics:

(1) by major types/subject coverage;

(2) by other aspects of categorization, namely
by purposes, i.e. to whom is information addressed? (clientele)
by input, i.e. what kind of data are processed?
by output, i.e. what kind of information is produced?
by geographical coverage, i.e. what area is concerned?
by language, i.e. what language(s) is/are used?
by the method of processing, i.e. what kind of categorization is used?
by organizational factor, i.e. what kind of institution is managing the system?

In its paragraph on the "Typology of international STI/SS" (scientific and technological information systems and services), the Freeman-paper[18] identifies and analyses three major types: *development-oriented, research-oriented* and *implementation-oriented* systems. A DEVSIS study, in turn, identifies two types: *discipline-oriented* and *mission-oriented* ones.[19]

The approach of the present study is different from the former: there are three major types of information systems, but the criteria for them are at variance with what has been proposed by the above-mentioned two papers.

Taking into consideration the aforegoing nine aspects of categorization, the present study proposes the following identification:

(1) *By major types/subject coverage*

It is possible to identify three main types: mission-oriented, sector/discipline-oriented, and normative systems/programmes.

(a) *Mission-oriented systems* are acting *horizontally*, covering all or most branches of activity, and of disciplines, irrespective of the interests of one or more particular sectors of socio-economic activities. Their objectives are *global, trans-sectoral* and *functional*, and are not branch-specific. UNBIS is one example for this type of system, covering all UN activities: politics, economy, law, science and technology, population, industry, and so forth, by controlling and indexing all UN documents concerned with the above. The UNEP's IRS (International Referral Service) is another case in point. In fact, what kind of activity is not concerned with and affected by the problems of environment? It is also the case for another two systems, the DEVSIS (Development Sciences Information System) and the SPINES (the UNESCO-initiated informations system for science and technology policies). They also aim at comprehensive coverage. If a population information system, concerned with human biology, public health, sociology, psychology, economics, environment and the like were set up, it would also be classed into this type of system. The specific feature of the mission-oriented systems does not merely consist in their inter-, multi-, or pluri-disciplinary character (this being only one aspect of their consideration). Practically, all sector/discipline-oriented or branch-specific systems are, to a certain extent, multi-disciplinary in character (e.g. information system on the food problem), however, the latter may be deemed as an activity oriented to one or more disciplines or sectors. This is not the case with any of the above-mentioned information systems, because neither the United Nations, nor the environment, nor development, nor science and technology are based on one particular discipline, nor are they representing only one or a few sectors of activity, bot are aimed at a *horizontal coverage* of international co-operation, embracing all (or the overwhelming majority of) types of branches, sectors of socio-economic activities. On this account, the most adequate term for this type of information system seems to be "mission-oriented".

(b) *Sector/discipline-oriented or branch-specific systems* are *vertical* ones, being concerned with a relatively well-specified socio-economic activity or with a discipline affecting one certain sector of public policy. In this respect, systems like AGRIS/FAO, INIS/IAEA, INDIS/UNIDO are all covering certain sectors or branches. In spite of their multifaceted subjects, they are sector/discipline-oriented systems in character. All of them are or can be based on several disciplines, at the same time they are concentrated on one main field of a well-defined sector such as agriculture, industry, and so forth. Unlike the mission-oriented system of horizontal character, these sectoral systems aim at a vertical coverage.

This is not to imply any judgement of value between the two major types of *operational* systems, the only difference between them lies in their respective basic orientation.

(c) *Normative systems/programmes* are non-operational in character as opposed to their mission-oriented or sector/discipline-oriented counterparts, instead they are acting on a normative basis. They are not providing information for any research "clientele", their "clientele" being composed of other information systems and library services. An appropriate example for this is the UNISIST which is an "effort to synthetize, at the international level, diverse philosophies, programmes, and policies related to the international exchange of information ...".[20] But UNISIST is also a policy "related to the international exchange of information" that initiates, within its framework, other systems like the ISDS with a normative element for the standardized identification of periodicals and serials. Normative systems have to play a very important role in the implementation of standardized methods, techniques, to make different systems compatible, thus promoting the harmonization and integration of information policy.

(2) *Other aspects of categorization*

(a) *By purposes* ("clientele"):
- government, policy and management-oriented (including development) (e.g. UNBIS)
- research-oriented (e.g. INIS)
- direct application-oriented (e.g. INDIS, INPADOC/WIPO patent documentation)
- service-oriented (e.g. INFOTERRA)

(b) *By input:*
- specialized literature (e.g. AGRIS)
- non-published material (e.g. INIS)
- official documents (e.g. UNBIS)

or by the combination of the three elements;

(c) *By output:*
- indices, bibliographies (e.g. UNBIS)
- abstracts (e.g. INIS, AGRIS)

or by the combination of these two elements;

(d) *By geographical coverage:*
- concerning a country
- concerning a region
- being on a world scale;

(e) *By language:* – monolingual
– multilingual;

(f) *By method of processing:* – traditional (manual)
– EDP-based;

(g) *By organizational factor:* – international organization (e.g. UNESCO/DARE)
– national organization (e.g. DEVSIS, Canada)

Information systems form a kind of "inventory management" of accumulated knowledge – to use Pierre Piganiol's expression.[21] "Such management requires a permanent inventory, followed by a classification which will never be definitive: knowledge must be regularly reclassified as new links are discovered between the various fields of science and technology." (p. 33, *op.cit.*). What holds true for classification, i.e. that it "will never be definitive", is also true for the categorization of information systems. With the change in their content (input), their dissemination of information (output), in the method of processing, etc., the category they are actually belonging to may also change. Therefore, the typology of information systems – like the systems themselves – is dynamic rather than static in character.

The role of the UNISIST programme and other systems

As has been stated above, UNISIST is a philosophy, a programme and policy for the co-operation of information systems on a world-wide scale. UNISIST is a successful attempt at bringing about a real and voluntary international co-operation, embracing the whole field of information activity, and – for the time being – it is the information-programme within the United Nations system which can serve as a catalyst for and a tool of an integrated policy.

Within its activity, supervised by a Steering Committee efforts of the UNISIST are concentrated on five directions as follows: improving the tools of interconnexion between the individual systems; improving information transfer; breeding a specialized information manpower; *developing science information policy* and national networks; *special assistance to developing countries*. Within this framework, UNISIST publishes guidelines, organizes training courses and seminars, etc. UNISIST has incorporated the formerly UNESCO/NATIS System. For current information on UNISIST activities see the *UNISIST Newsletter* published by the PGI/UNESCO.

Despite all this, UNISIST does not act as, and cannot replace an over-all United Nations information policy, but serves as a catalytic "master programme" and a normative "umbrella" for information systems; it can be a bridge, and intercommunication among systems within and without the United Nations. It is a powerful means of co-ordination in a United Nations information policy.

It appears from the typology of the United Nations information system that all the major fields of socio-economic activities *are supplied* with information by the established systems and services.

What really counts is not the number, small or large, of the United Nations information systems but their *fundamental interrelations* and the rational division of labour among them.

All this is involved in policy and management which form the subject of the following chapter.

III. How is information managed by the United Nations?

Management, in fact, must follow from a policy. For the time being, however, no information policy exists for the United Nations system as a whole, or more precisely, what exists is a quasi-policy, and consequently a quasi-management of information. Any leading body related to development and to the socio-economic activities of the Organization, has no "mandate" to formulate or conduct a policy for the United Nations family on information and its management. There are committees, working groups, task forces etc. for co-ordination the information policies of the UN family, but there does not exist any high-level body or office (department, division) with authority which is directly responsible for tackling the problems of information. Many years after having been formulated, the statement in the *Jackson-report* is still valid: "... each of the organic services of the United Nations, and some specialized institutions, plans and administers its own programmes without worrying particularly about the programmes of other bodies."[22]

The highest body for policy, management and co-ordination in the United Nations system is the Administrative Committee on Co-ordination (ACC) reporting to the Economic and Social Council. A subsidiary body of the ACC is the Inter-Organization Board for Information and related activities. But even if it were possible to consider ACC/IOB as a leading body for information, an important potential group of information institutions would not fit into the picture: the United Nations *libraries* whose activities are not even formally co-ordinated.

A real power to co-ordinate consists in *financing power*. Such a co-ordinating body doesn't exist in UN system for information and libraries. Another consideration, which is connected with the former, lies in the fact that attempts at information policy are not properly linked to the social-economic development activities, to the R&D policy of the UN.

Policy, decision-making and management of information within the United Nations has to be concentrated, then integrated with the development policy, finally co-ordinated in conjunction with financial decision-making power. National experiences show also that without such a *concentration, integration* and *financing*, any attempt at co-ordination remains wishful thinking.

The scheme of the existing situation is self-explanatory: decision-making, financing, management on the one hand, supervising, reporting on the other are separated, and the United Nations libraries are not connected with the other information services.

Special mention can be made of the lack of policy on the United Nations libraries. In 1946, the Economic and Social Council designated the Headquarters Library to act as adviser in the matter of library policy, but it has not been specified what this responsibility consists of. Except for a few technical actions, e.g. standardization of microfiches, those concerned with the depository libraries have done very limited contribution for a substantial common work.

However, from the ACC reports on co-ordination concerning information systems and the IOB, it appears that an impressive number of decisions, recommendations, etc. have been prepared in the field of information.[23] Between 1961 and 1973 not less than 30 such items are listed (General Assembly, ECOSOC) on information policy. In fact, the number of decisions, recommendations, official reports, studies of the General Assembly, ECOSOC, UNDP, etc. on the "information support" only in the period 1946–1975 can be estimated at 350 to 400 items. To be found among them are such important studies as the mentioned Jackson-report or the Henderson-study.[24] The same applies to the volume of resolutions, projects, studies, etc. concerning "information support" by the specialized agencies, with special regard to UNESCO; after having launched the UNISIST programme, NATIS was launched four years later as another independent programme. (See under PGI's UNISIST Programme.) Other services and systems initiated by UNESCO are: CDS (Computerized Documentation Service – operational), DARE (Data Retrieval for Social Sciences), ISORID (for current research in documentation – operational), SPINES (acting by limited scale as a science and technology policy information exchange system) – to quote only a few examples for the abundance of initiatives concerning "information support".

As to initiatives taken within the United Nations, they remind us of Baudelaire who, in his famous Albatros, says: "Ses ailes de géant l'empechent de marcher". And as the Duke puts it in Shakespeare's Othello (Act I, scene III.): "To vouch this is no proof", in other terms, the volume of initiatives by itself is not sufficient "without more wides and more over tests" (Shakespeare: *op. cit.*), i.e. without an effective management of information transfer.

A. A simplified scheme of the information policy and management and financing by the United Nations

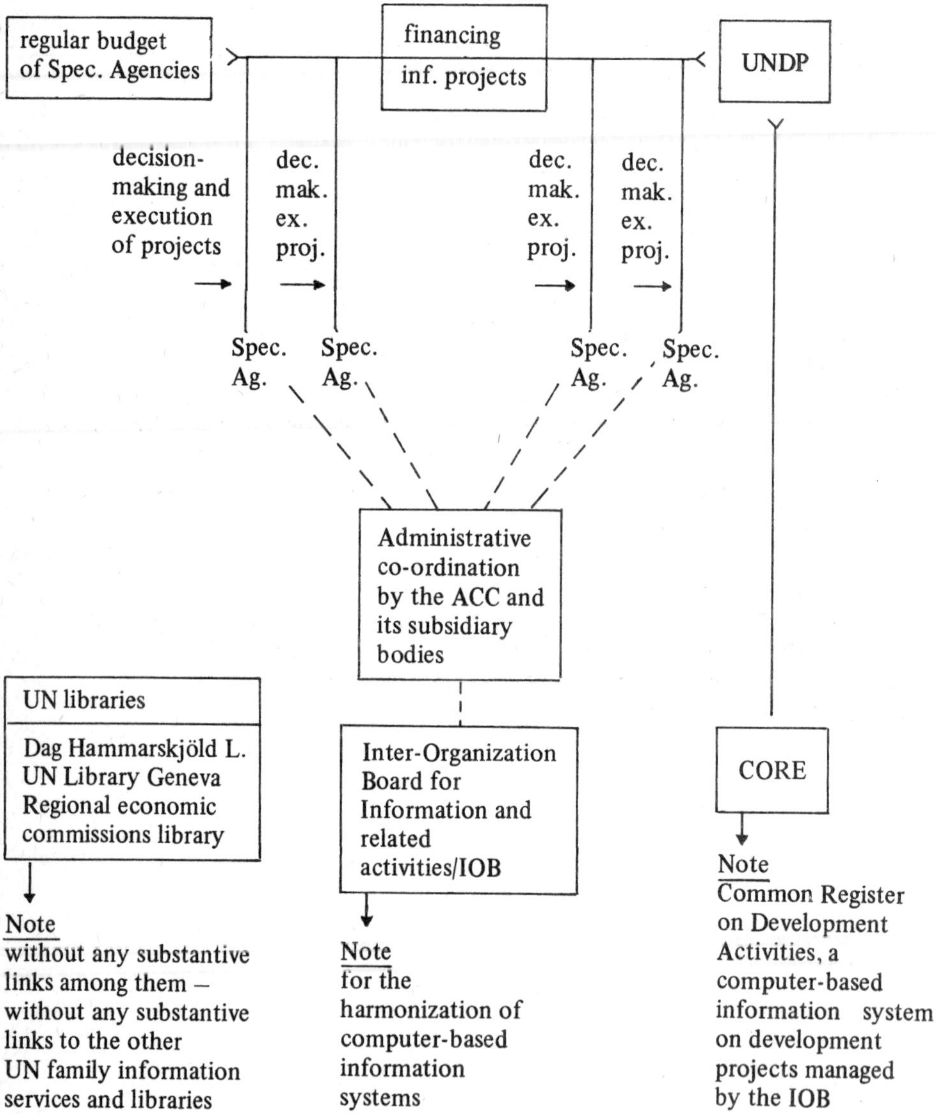

——— decision-making, management and financing information projects

- - - - co-ordination links

ACC Administrative Co-ordination Committee, reporting to the ECOSOC.

IV. An attempt to integrate U.N. information systems into a real network

It appears from the previous chapter that the present status of information within the United Nations is like producing goods (information services) of high quality materials which are provided plentifully by the existing and proposed information systems or programmes, but they are not well-proportioned to the needs of development.

Steps to change the present situation and the leading role of the United Nations in this process

The *first step* to change the above-mentioned situation is the unambiguous consideration of the transfer of information as a problem of development policy rather than that of administrative co-ordination.

Secondly, the consideration of information policy as an integral part of science and technology policy.

Thirdly, as a consequence of the above two considerations, to be developed as an integrated policy and management of information transfer for the whole United Nations system.

Following from the former and as the *fourth step*, it is necessary to transform the existing quasi-network, whose members are alive but not the network itself, into a real and efficient complex of systems for the transfer of information.

As the *fifth step*, decision-making, financing by UNDP, as well as the supervision of management and reporting has to be changed to form an integrated and interconnected process.

From the institutional point of view, all the existing bodies may be considered for such a change, along with their experiences, positive or negative. For the time being, UNISIST has proved to be the most successful programme of co-ordination. It forms an "umbrella" for and a "bridge" between the United Nations and various other international organizations and national information activities. But UNISIST is essentially a normative programme rather than an over-all United Nations information policy. It may be an important tool or programme of such a policy, but cannot replace it. UNESCO itself as a specialized agency for education, science and culture has accumulated highly important experiences in the field of information, but from the viewpoint of information transfer, the really efficient partners seem to be the over-all – and not the sectoral – decision-making bodies of the member States like ministries for development, planning offices, and similar bodies whose partners are the United Nations, the Economic and Social Council, and the United Nations specialized bodies for development. The IOB is more a technical co-ordination body, but much in the same way as the UNISIST it can play a supporting role in an integrated United Nations information policy.

As has been discussed, the United Nations has to play the *leading role* in the policy and management of information transfer, integrating it with its development policy, for two reasons. First: the United Nations is the most adequate partner for the member States in the matter of development; secondly: only the United Nations can have the authority necessary to an integrated policy and management within the United Nations family.

How to change?

The integration of the transfer of information with the development policy involves changes both as regards management and the institutions.

"The scheme of an integrated United Nations information system" demonstrates the main orientations of these changes in the flow of information, in organizational links and in the standardization of methods and techniques. As to the details, four additional schemes will show what is imagined, including computer-based data processing. The related orientations are as follows in the order of priority:

(1) *Intergovernmental Committee.* There are two alternatives, (a) the creation of a new Committee on Information Planning and Management for Development (CIPMD) to be the leading body of the United Nations family for the policy and management for information transfer with a status similar to CSTD (Committee on Science and Technology for Development) or CDP, reporting also to the Economic and Social Council, and (b) to assign this function to CSTD or to the Committee on Development Planning.

(2) *A Co-ordinating Unit in the Secretariat.* Again there are two alternatives, (a) either the creation of a special Office for the Transfer of Information for Development (OTID) to act as an executive organ of the over-all information policy within the Department of Economic and Social Affairs, having the same status as OST or (b) the inclusion of the function of such an office into the terms of reference of the United Nations Office for Science and Technology (OST).

(3) For the more rational and economical computer-based processing of data for users, to be developed is a central unit, the COPRO (*Co*mmon *Pro*cessing or *Co*mputer *Pro*cessed Data or Inter-Agency *Co-pro*duction of EDP data).

(4) The normative catalyst role of the *UNISIST* has to be strengthened and closely connected to the OTID.

(5) The OIB representing the managerial skill for the EDP-system has to be connected to UNISIST, and has to act for or be transformed into *COPRO*.

B. Scheme of an integrated United Nations information system

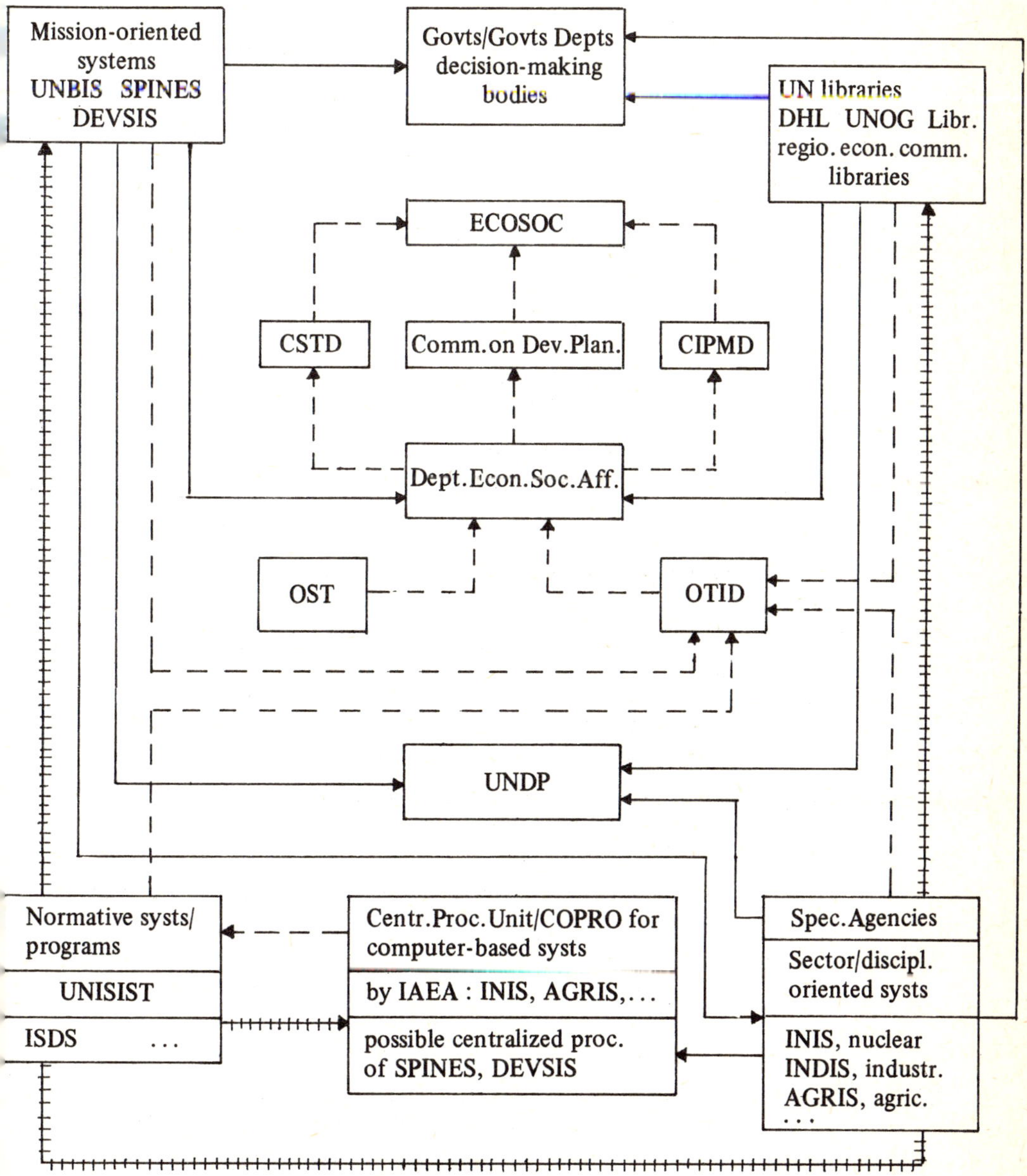

——→ the flow of information, from . . . to

– – – –→ the organizational links, reporting . . . to

++++++→ the standardization of methods/techniques, methods . . . applied by

C. Scheme of the flow of information

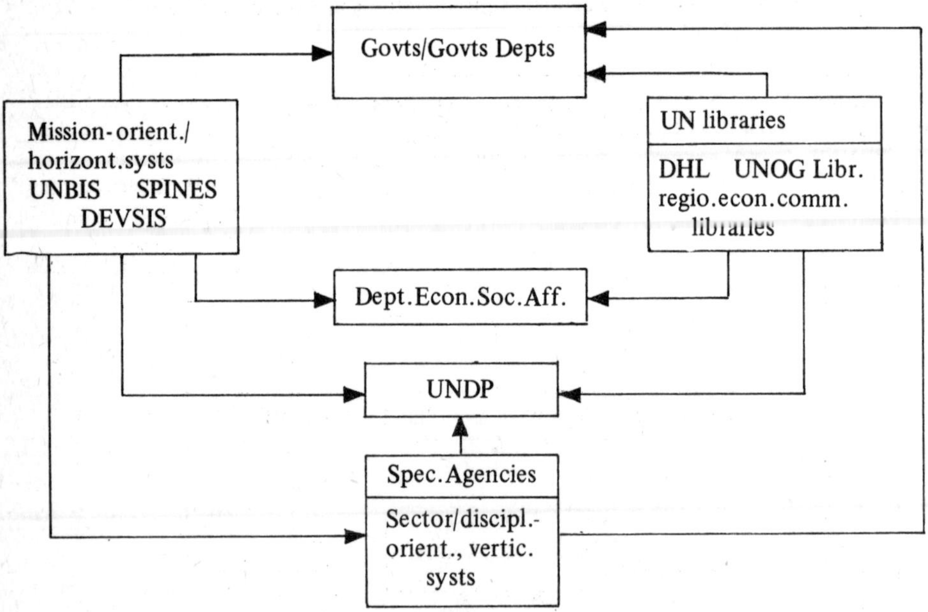

The *main users of information* supplied by the United Nations information systems and services are as follows:

Governments receiving information from the *mission-oriented* and *sector/discipline-oriented systems*, and from the proper *United Nations libraries and information services* (these are: the Dag Hammarskjöld Library, the UNOG Library, and the libraries of the regional economic commissions);

United Nations Department of Economic and Social Affairs – receiving information from the *mission-oriented systems* and the *United Nations Libraries;*

UNDP – receiving information from the *mission-oriented* and *sector/discipline-oriented systems*;

United Nations specialized agencies – receiving information from their own *sector/discipline-oriented* and *mission-oriented systems*. (Specialized agencies are both the producers and users of information);

The main *producers of information* are the following:

Mission-oriented systems – providing information for *Governments, Specialized agencies, UNDP, Department of Economic and Social Affairs.*

Sector-discipline-oriented systems – providing information for *Governments*, for *themselves*, and for the *UNDP*;

United Nations libraries – providing information for *Governments*, for the *Department of Economic and Social Affaits*, and the UNDP.

D. Scheme of organizational links

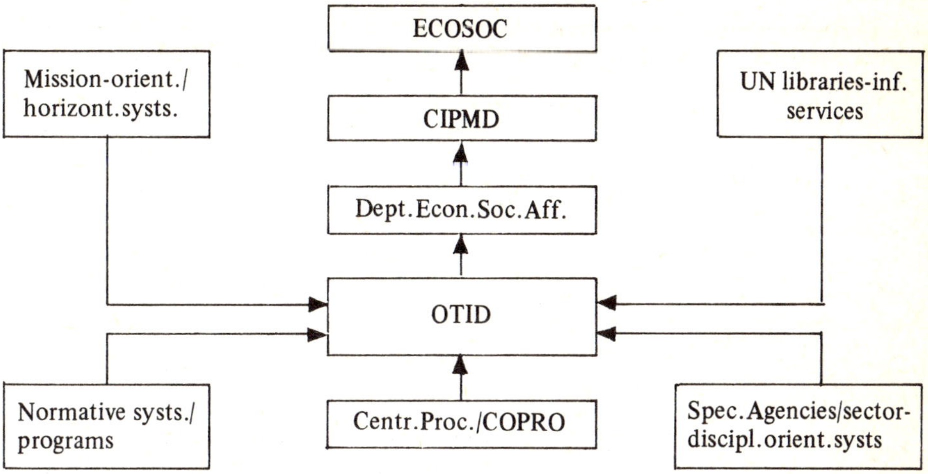

Within the United Nations system, the new *OTID is mainly concerned* with the *policy* and *management* of information transfer, under the overall supervision of the *Department of Economic and Social Affairs* and under the guidance of the new *Committee on Information Planning and Management for Development*, reporting to *ECOSOC*. The following information systems and services are guided *by* and reporting *to*:

OTID – *mission-oriented systems, normative systems, COPRO* (central EDP unit), *sector-discipline-oriented systems* (which are also reporting to their governing boards), *United Nations libraries*.

The OTID, jointly with the UNDP, assumes the task of decision-making, manages and controls the funds of the Development Programme allocated for information.

The CIPMD supervises and reports to ECOSOC on the policy, management and use of regular and special budgets for information of all the United Nations-related organs, including the specialized agencies.

E. Scheme of standardization of methods/techniques

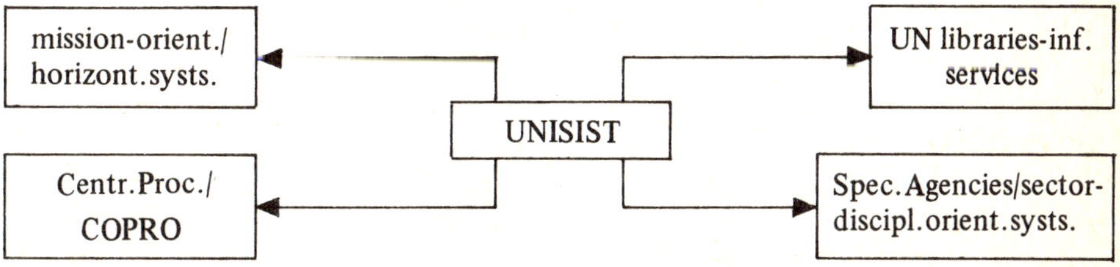

The focus of the standardization of methods/techniques in information is the UNISIST as the main normative system:

UNISIST – orienting the *mission-oriented systems*, the *COPRO* (central EDP unit), *sector/discipline-oriented systems*, and the *United Nations libraries*.

F. Scheme of common computer-based processing of data

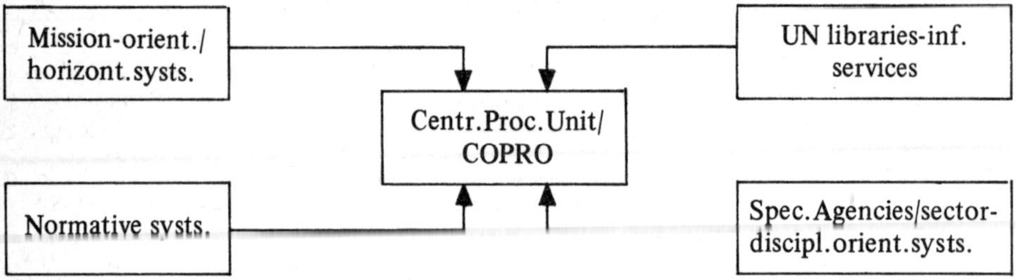

The *central processing* of *computer-based* information systems can be performed by the COPRO (*Co*mmon *Pro*cessing of computer-based data or *Copro*duction of EDP-based information):

COPRO – the processing of data of the *mission-oriented, normative* and *sector/-discipline-oriented systems*, as well as those of the *United Nations libraries*. The centralization of the EDP-based information does not mean that all systems should be necessarily processed by COPRO; this is the question of a cost/benefit analysis, and that of the rational utilization of hardware and software, and does not exclude the possibility of the *particular* processing of information where it is justified.

V. A new approach to information transfer for developing countries: the Applied Information Centres – AIC

It seems that a change in the policy and management of the transfer of technology, the integration of information policy with the development policy within the United Nations and the concentration of information activities can be important steps, but all this represents only one side of the problem. The other and not less important side of the problem is the *power of adaptation* of the transfer of information of developing countries, in other terms, the *information infrastructure*.[25]

Applied Information Centres as "focal points" of information transfer in developing countries

Information activities tend to proliferate not only within the United Nations, but also in developing countries. This phenomenon is closely interrelated on both sides. Each information system and programme is inclined to organize its own local "focal point", its "centre of application". There are focal points, like those for national information policy created under the auspices of UNISIST and those for IRS/UNEP; other networks of focal points are being established or envisaged for DEVSIS, SPINES etc. However, there will be a certain moment when it will be necessary to organize a "focal point" of "focal points" in developing countries, and this will pose some problems. This is the *organizational* aspect of "focal points".

But it is much more important that the *substantive* aspect of the adaptation of information transfer which – generally speaking – follows the classical model of industrialized countries is based on high specialization. Information systems by sector (agriculture, industry), computer-based systems, libraries, training, education, etc., *sector by sector*, without, however, being connected among themselves. With a view to the lack of skilled manpower, qualified professionals, the *training* and the *human factor* should be given *absolute priority*. Therefore, a *trans-sectoral* approach to this problem has to be envisaged.

The trans-sectoral approach means that a new kind of organ can be developed, based not on the advanced specialization, but on the *organic interconnexion* of information-training-extension. This can solve, at least, the more rational and economic utilization of manpower, and may open up a new way for the effective management of information transfer *to* and *by* developing countries.

The prototype of an Applied Information Centre (AIC)

"Information-by-doing" clearing house: a complex of information-training/recycling by working, equipped with the techniques of agricultural (and/or other sectoral) extension services and productivity centres.

Moduli of AICs

(a) *Reference Service*

Selected collections of reference works, textbooks, guides, "how-to-find" lists, repertories, collective catalogues, encyclopaedias, new issues of relevant serials and journals.

(b) *Automated Information Service*

Display terminal connected to the big teledocumentation systems, equipped with teletypes and rapid copying machines.

(c) *Audio Visual Documentation Service*

Films, recordings, exhibits and other extension methods and techniques.

(d) *"Know-how" Service*

Workshops for agricultural and industrial training, experimental stations, and other means and techniques of production for adult training and refresher courses.

(e) *Conference and Consultations Service*

Preparation of courses, lectures, combined with demonstrations (see under (c)) and with production (see under (d)), to provide agricultural, marketing and industrial "know-how" and pedagogic materials, to diffuse guides, curricula, to give written or oral counsel and advice.

(f) *General Services*

Management, staffing, finance of AICs, including laboratory for applied psychology and professional orientation, and a unit for economic and socio-cultural analysis; the GS at the same time could be a point for the training of the future specialists in administration and management.

The moduli are inter-changeable according to the size and priorities of the AICs. For instance, the Reference Service could be combined with the Automated Information Service, the Audio-Visual Service could be attached to the "Know-how" Service, and there might be several other forms of combination.

The AICs are complex units under one single management, including different aspects of technical assistance: training, professional education, information of various sectors like agriculture, industry and so forth. The uniform management of the AICs reflects the integrated information policy for development by the United Nations, thus creating from them *real "focal points" of the adaptation of information transfer*, and through the latter, "focal points" for the transfer of technology which is the real sense of information activities *of* and *for* developing countries.[26]

Computer techniques: not a cure-all

There arises the question of why AICs and not the computer-based information services are suggested to be developed in and for developing countries as soon as possible? The answer is that the application of the highly sophisticated techniques and methods is useful only in case if the infrastructure and – above all – the organizational and human factors in developing countries are ripe for the use of such techniques.[27]

The idea of the information explosion and information revolution with the *computer as a panacea* has influenced, to a certain extent, the developing countries, and has added some confusion to the whole problem and led to a certain dissimulation of the real questions of information for development.

However, it is not the computer in itself that is expensive, but *the illusion* that with it, everything will be better. This diverts energies from their real object and puts them on the wrong track. The misunderstanding concerning the automation leads to waste, as it always happens when the means employed are clearly superior to the conditions in which they are or will be used. This means that the use of sophisticated techniques is justified only if the moment is well chosen, which is not often the case. It is not the computer that brings about development, but at a certain stage of expansion, an economic and technological infrastructure may require a changeover to a higher technical level. Conversely, if the moment of this changeover is not well chosen, technological dependence will be even stronger instead of being reduced.

There is one thing that must be kept in mind with regard to computerized information – its *impersonality* and the fact that it deprives the user of reading. The reading of specialized literature is a means of training and recycling. Therefore, library services have to be promoted first for the *advancement of reading*.

Every country has to be responsible for its information policy taking into account international experience and especially the experience of countries under the same conditions.

Generally speaking, information for development has to give absolute priority to social and human factors. Various things here can be implemented in this respect. Among these the system of T+V (Training and Visit) implemented by World Bank specialists is to be quoted. This is an example of a combination of the advising/training and information activities all based on personal contacts.[28]

Pilot studies for AICs and the role of the United Nations regional economic commissions

The implementation of the AICs has to go step by step, relying on experiments and taking into account the different levels of development of various regions, the specific economic and socio-anthropological conditions of developing countries.

It seems, therefore, to be appropriate to conduct pilot studies by teams composed of economists (planners, managers, etc.), cultural anthropologists and information specialists in certain regions of Latin America, Africa and Asia in order to gain the know-how in implementing the AICs. At the same time these teams could evaluate the present situation of the concrete information activities for and in certain countries (regions) and use their experiences for their recommendations. On creating the AICs these pilot studies would be inter-agency and trans-sectoral in character. The trans-sectoral and inter-agency character means that the functioning of the AICS cannot depend on one sector of socio-economic activity and cannot take into account only one or two specialized agencies of the United Nations but it must respect the integrated information network of the United Nations and its specialized agencies.

As regards the complexity of the AICs functions – "information-by-doing" training and extension – the following agencies could be interested in it: ILO as an agency for the promotion of training, UNESCO/UNISIST as the agency for education and science and as a catalyst organ for all information activities, and – depending on the economic conditions – the branch specific agencies like FAO for agriculture, UNIDO for industry should also be involved. The UNDP as the financing body of all development projects should also be interested in it. On account of the diversity of institutions involved, it seems to be useful to introduce an independent management of the AICs under the guidance of the United Nations Secretariat. The ACCIS (Advisory Committee for the Co-ordination of Information Systems) has to plan a major role in searching new ways, means and methods in developing the developmental information and infrastructure.[29]

This independent management can be ensured by the *regional economic commissions* which are *near* the AICs and are directly involved, having the competence in socio-economic problems in the concerned area. Generally speaking, the regional economic commissions have to assume responsibility directly for information for development by extending their duties and activities in this matter. This could be the most rational and economic way for the users of the transfer of technology. The regional economic commissions could become powerful organs of the effective management of the transfer of technology. All things considered, the objective of the transfer of technology is to reduce dependence on technology.

G. A scheme for the setting up of Applied Information Centres

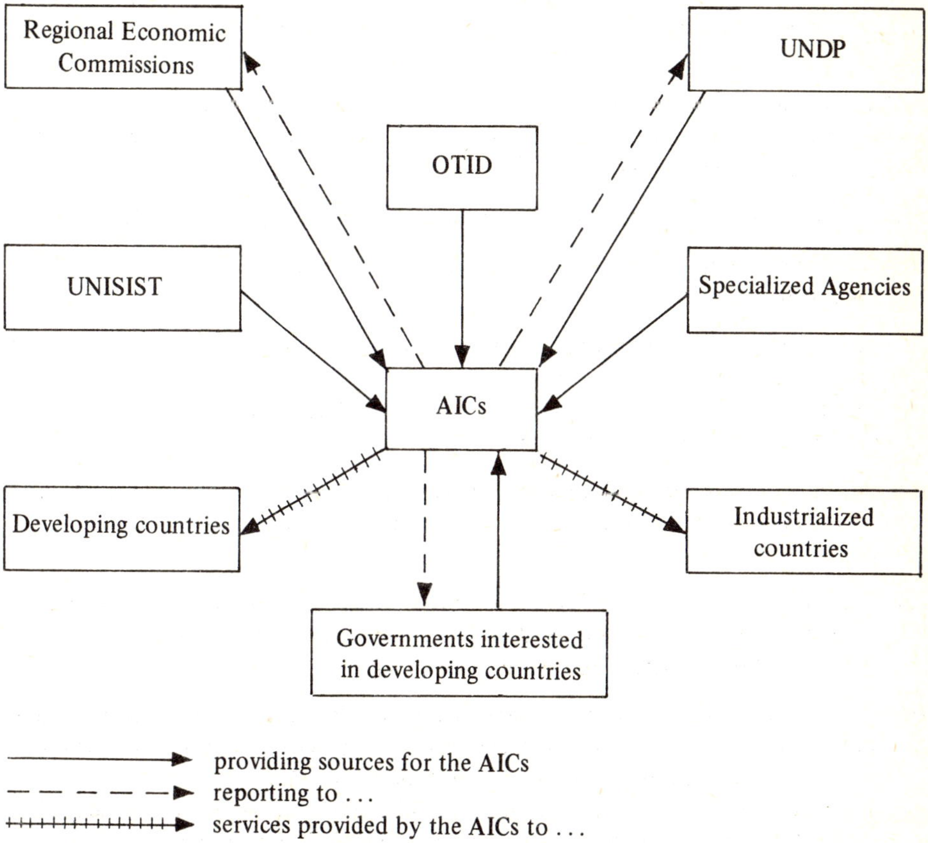

OTID = Office of Transfer of Information for Development (as a new UN Secretariat Unit)

AICs = Applied Information Centres

VI. Conclusions

(a) A general revision of the technical assistance in the matter of information is not only in the interest of developing countries, but also serves the interests of both the industrialized countries and the United Nations system.

(b) The development information policy and management of the United Nations has to be an integrated part of the general economic and social policy, as well as of the scientific and technological policy.

(c) There is no general model for providing and improving development information.

(d) Developing countries themselves can solve their information problems mainly by their own efforts.

(e) In contrast with industrialized countries where information services are the result of the affluency of data, developing countries are confronted with the lack of relevant data on the one hand, and with the uncoordinated flow of data provided by the technical assistance, on the other.

(f) Information activities in developing countries has to be based on their actual socio-economic infrastructures, taking into account the specific cultural-anthropological status.

(g) The implementation of EDP-based information services in developing countries has to be preceded by the institutional maturity of their infrastructures: priority should be given to human factors, to training and management.

(h) Researches guided by complex teams of economists (planners, managers), cultural anthropologists and information specialists are to be conducted to find new types, means and methods of the transfer of information for development; an example of a new type of institution for the transfer of information can be the Applied Information Centres – a kind of "information-by-doing" clearing house.

NOTES

1. *UN Economic and Social Council.* Resolution 1902 adopted at the fifty seventh session (1 August 1974) on the role of an international technological information system in the transfer and assessment of technology, and in the indigenous growth of appropriate technologies in developing countries.

2. *Kennerley, J.A.:* The role of an international technological information system in the transfer and assessment of technology, and in the indigenous growth of appropriate technologies in developing countries, UN Office of Science and Technology, New Technologies Section, 11 October 1974. Mimeo.

3. *Freeman, J.E.:* Improving technological information exchange with developing countries... a feasibility study, Phase I, report. UN Office of Science and Technology. February 1975. 41 p.

4. *UN Advisory Committee on the Application of Science and Technology to Development.* Twenty first session, Geneva, 17–28 February 1975, E/AC.52/XXI/CRP.8. (See 1/ and 2/.)

5. UN, General Assembly, Seventh Special Session, A/RES/3362/(S–VII), 19 September 1975. Resolution... Development and international economic co-operation.

6. UN, General Assembly, Seventh Special Session, A/10232, 24 September 1975. Report of the Ad Hoc Committee... Development and international economic co-operation.

7. UN, General Assembly, Sixth Special Session, 3203(S–VI), 1 May 1974. Programme of action on the establishment of a new international economic order. (See Chapter IV. Transfer of Technology.)

8. UN, General Assembly, Sixth Special Session, 3201(S–VI), 1 May 1974. Declaration on the establishment of a new international economic order. UN, General Assembly, Resolution 2626(XCV).

9. UN, A new UN structure for global economic co-operation. E/AC.62/9 1975. (UN Publication, Sales No. E.75.II.A.7)

10. ACAST = Advisory Committee on the Application of Science and Technology to Development, the same organ which – through the UN Office of S and T – had commissioned the author's original study. The present paper is based on that.

11. *UNESCO,* UNISIST, Intergovernmental Conference for the Establishment of a World Science Information System, Paris 4–8 October 1971. Final Report. Paris 1971, SC/MD/25.
UNESCO, UNISIST, Study report on the Feasibility of a World Science Information System, Paris, 1971, SC/70/D.75/A.
UNESCO, Science and Technology Policies Information Exchange System SPINES. Feasibility Study. Paris, 1974, Science Policy Studies and Documents. No. 33(1).
The whole documentation of the three 1979 Vienna-Meetings is to be found in the Dag Hammerskjöld Library (UN New York) and at the UN Office at the Geneva Library.

12. ACCIS Newsletter (Advisory Committee for the Co-ordination of Information Systems, ISSN 0254–3133), Geneva UN, 1981 – and
Update. UN Centre for Science and Technology for Development. A quarterly publication of UNCSTD, distributed by the Communication Cluster of UNCSTD, UN Secretariat, New York.

13. *UNESCO,* PGI – General Information Programme and UNISIST. Paris, 1984. 16 p.

14. *ACCIS Newsletter* 5(1) May 1987 Supplement:
Indexing Vocabularies Produced by Organizations of the United Nations System
– an abridged version – (in the original 40 items)

United Nations (UN)
1. Macrothesaurus: for information processing in the field of economic and social development / United Nations Department of International Economic and Social Affairs. – 3rd ed. – New York : UN, 1985
xiv, 347 p.
2. UNBIS thesaurus: Trilingual list (English, French, Spanish) of terms used in subject analysis of documents and other materials relevant to United Nations programmes and activities / Dag Hammarskjöld Library. – English ed. – New York : UN, 1985
xxiv, 557 p.

United Nations Environment Programme (UNEP)
3. INFOTERRA : thesaurus of environmental terms. / Nairobi : UNEP, 1984
66 p.

United Nations Centre for Human Settlements (Habitat)
4. UNCHS (Habitat) trilingual thesaurus : in the field of human settlements / UN Center for Human Settlements. – Nairobi : UNCHS(Habitat), 1986
126 p.

United Nations Children's Fund (UNICEF)
5. UNICEF thesaurus / United Nations Children's Fund. – New York : UNICEF, 1985
125 p.

United Nations Development Programme (UNDP)
6. UNDP thesaurus / United Nations Development Programme. – Draft ed. – New York : UNDP, 198?

Office of the United Nations High Commissioner for Refugees (UNHCR)
7. Thesaurus of refugee terminology / compiled by Piers Campbell. – Prelim. ed. – Geneva : Refugee Documentation Centre, United Nations High Commissioner for Refugees, 1986
vi, 287 p.

United Nations Fund for Population Activities (UNFPA)
8. POPIN thesaurus : population multilingual thesaurus = thesaurus multilingue de population = tesauro multilingue sobre población / prepared by Jean Viet for Population Information Network Committee for International Information Network Committee for International Co-operation in National Research in Demography, United Nations Fund for Population Activities. – 2nd ed. – Paris : CICNRD; New York : UNFPA, 1985
3 v.

International Labour Office (ILO)
9. ILO thesaurus : labour, employment and training terminology = Thesaurus BIT : terminologie du travail de l'emploi et de la formation = Tesauro OIT : terminologia de trabajo empleo y formación. – 3rd ed. – Geneva : ILO, 1985
xxvii; 490 p.

Food and Agriculture Organization of the United Nations (FAO)
10. AGRIS/CARIS categorization scheme. – 4th ed. – Rome : FAO, 1983.
161 p.
11. AGROVOC : a multilingual thesaurus of agricultural terminology = thesaurus multilingue de terminologie agricole = tesauro multilingue de terminologia agricola / Food and Agriculture Organization of the United Nations and Commission of the European Communities. – Rome : Apimondia, 1982.
7 v.

United Nations Educational, Scientific and Cultural Organization (UNESCO)
12. DARE permuted descriptor list. – Paris : UNESCO, 1987.
Computer printout.
13. International thesaurus of cultural development = thesaurus international du développement culturel / prepared by Jean Viet. -- Paris : UNESCO, 1980.
xl, 498 p.
14. SPINES thesaurus : a controlled and structured vocabulary of science and technology for policy-making, management and development / compiled by UNESCO Secretariat and B. de Padirac. – Paris : UNESCO, 1976.
3 v. (623 p.)
15. Thesaurus mass communication = thesaurus communication de masse = tesauro communicación de masas / by Jean Viet. – 2nd ed. – Paris : UNESCO, 1982.
xviii, 356 p.
16. UNESCO thesaurus : a structured list of descriptors for indexing and retrieving literature in the fields of education, science, social science, culture and communication / compiled by Jean Aitchison. – Paris : UNESCO, 1977.
2 v.
17. UNESCO-IBE education thesaurus : a faceted list of terms for indexing and retrieving documents / International Bureau of Education. 4th ed. – Paris : UNESCO; Geneva : IBE, 1984.
361 p. ISBN 92–3–102061–7

International Monetary Fund (IMF)
World Bank
18. The World Bank thesaurus / World Bank and IFC Records Management Division, Joint Bank-Fund Library. – 2nd draft ed. – Washington, 1984.
2 v. (ix, 1040 p.)

International Atomic Energy Agency (IAEA)
19. INIS thesaurus. – 25th ed. – Vienna :IAEA, 1986
813 p.

United Nations Industrial Development Organization (UNIDO)
20. Thesaurus of industrial development terms : English-French / United Nations Industrial Development Organization. – New York : UN, 1983.
ix, 152 p.

International Trade Centre UNCTAD/GATT (ITC)
21. Thesaurus of international trade terms. – Geneva : International Trade Centre UNCTAD/GATT, 1982.
163 p.

Inter-Organization Board for Information Systems (IOB)
22. Board terms for United Nations programmes and activities / prepared by the Inter-Organization Board for Information Systems. – Geneva : IOB, 1979 (no revisions)
viii, 186 p.

15. *Myrdal, G.*: What is development? – Ekistics, August 1975. No. 237. pp. 84–87.

16. A few important publications in the field of international information systems:
Directory of United Nations Information Systems. Vol. 1–2. Geneva, 1980. UN. 465, 215 p.
Directory of United Nations Databases and Information Systems 1984. New York, 1985. UN. 323 p.

Documents of international organizations: a bibliographic handbook covering the United Nations and other intergovernmental organizations. Comp. and ed. by *Th. D. Dimitrov*, Chicago, 1973, American Library Association. 301 p. – This is a fundamental reference work in the field.

International organizations. United Nations. Comp. and ed. *Th. Dimitrov*, 1982. Berlin, de Gruyter. 41–119. p.

Marully, L. Documentation of the United Nations system. Metuchen, 1979. 225 p.

Organization for Economic Co-operation and Development. Information activities of major international organizations. Paris, 1971. 175 p.

Sources, organizations, utilization of international documentation. Proceedings of the International Symposium on the Documentation of the United Nations and other Intergovernmental Organizations, Geneva, 21–23 August 1972. The Hague, 1974. FID publ. 506., 586 p. – A basic work on international systems and programmes.

Tocatlian, J.: International information systems. *Advances in Librarianship*, Vol. 5. 1975. pp. 2–52.

United Nations, Economic Commission for Europe. Inventory of major information systems and services in science and technology. Geneva, 1971. 340 p.

17. *Ibid. 14/*

18. *Ibid. 3/* pp. 9–10.

19. DEVSIS: Preliminary design of an international system for the development sciences. Geneva. ILO/DEVSIS Study Team – 1975. 258 p.

10. *Ibid. 11/* UNESCO/UNISIST *Piganiol, P.*: Science and information in prospect.

21. *Ibid. 11/* Appendix B.

22. *Jackson, R. G.*: A study of the capacity of the United Nations development system. Vols. 1–2. Geneva, 1969. (United Nations. Documents. DP/5.) United Nations Publ. sales No. 1970. I. 10. See in its vol. 2., chapter VI. Information system concept.

23. *United Nations*, ECOSOC, Twenty-seventh session, Co-ordination reports of the Administrative Committee on Co-ordination. Information systems, Special report of the ACC and Addendum: An analysis of United Nations initiatives relating to information support. 9 May 1974. E/5489 and E/5489/Add. 1.

24. *Ibid. 22/* and *Henderson, A, M.*: Electronic data processing in the United Nations family of organizations: a concept for effective growth and utilization. Vols. 1–2. New York, 1970. (United Nations, General Assembly, twenty-fifth session. Documents. A/8072.)

25. For the problems of information infrastructures in developing countries, including the AICs (Information Applied Centres) as new organs the author of this paper had written a few articles and studies, let us quote the following ones: UNESCO. Republic of Maldives. Development project for the National Library. Report prepared for the Government... Technical Report. PP/1981–1983/5/10.1/05. Serial No. FMR/PGI/82/179. (Rózsa) Paris, 1983. UNESCO 17 p. UNESCO. Rwanda. Le développement d'un systeme intégré d'information scientifique (BACIN). Rapport de mission PP/1981–1983. No. de série FMR/PGI/83/189. (Rózsa) Paris, 1983. UNESCO. 21 p. The great illusion. *CERES* (FAO review for development). September–October 1975, pp. 35–38. (In this book see 185. p.)

26. *Ibid. 25*/ see: UNESCO. Rwanda...

27. *Ibid. 25*/ see: The great illusion...

28. *Fehr, B.:* Das erfolgreiche System des Daniel Benor. Frankfurter Allgemeine Zeitung. 3 Januar 1987. Annex.

29. ACCIS was established on 31 March 1983 by decision of the Administrative Committee on Coordination (ACC) in response to resolution 1982/71 of the UN/ECOSOC. ACCIS facilitates access by Member States to UN information and promotes the improvement of the information infrastructure within the UN system. Each organization of the UN system has designated a focal point which is responsible for publicizing ACCIS activities and involving its organization in relevant ACCIS projects. ACCIS is managed by a Steering Committee and its Register of Development Activities is in a planning stage. In Dec. 1984, ACCIS published the *Directory of United Nations Databases and Information Systems.* This is a guide to 615 selected information systems, information services and databases of the United Nations system.

MALDIVES CHRONICLE

(This account was written in 1982 at the time of my mission and it refers to a particular experience. Although the conditions may have been modified in the intervening years, there is, I feel, some sort of a lesson of general substance regarding developing countries, for which reason this chronicle might still be valid.)

The Maldives are a group of islands in the Indian Ocean, in the equatorial region, equidistant from Kerala and from Sri Lanka. They are in the neighbourhood of the more famous Diego Garcia (U.S. base) and the Seychelles which appear on the brochures as a paradise. For the tourist the Maldives could also appear to be a paradise: excellent climate, beautiful vegetation, exotic geographical position, maximal possibilities for relaxation – and all that quite cheaply, provided one gets there first from Europe which is 6000 miles away.

The Maldive islands are an atoll republic. Atoll comes from Divehi, the national language of the republic. It is Arabic in origin. The atolls consist of oval shaped coral islands which rise out of the sea gently and form straights with tranquille internal harbours. The total area is about 60,000 m^2 covering water and land with twenty atolls, meaning groups of islands. There are about 1200 islands, most uninhabited, but the 200 of them which have a population, add up to about 150,000. Out of these 40,000 live in the capital Male, which belongs to the North Male atoll and is very densely populated. People live in houses and in junkas, which are houseboats but often use and live in fishing vessels as well.

Should there be a runway or should the airport be in the capital – these questions must have been posed to the planners. At the end, Hulele airport was built 15 minutes away from the capital when travelling on a motor barge. One feels one has really not *landed.*

What sort of civilisation?

Why did I go there? In order to write a report to the U.N. concerning reading, information, libraries. Why me? Perhaps because I had worked in the U.N. in Geneva and they have remembered me. From 1888 to 1964 the predecessor of the Republic was a British Protectorate, which, in 1965, was granted independence. Once the overseer lived in Colombo, the power was in the hands of the local sultan and the people confessed to the Sumnite variety of Muslim faith. There is a curious ambivalence: economically speaking the Republic is among the poorest 25 countries in the world, yet basic literacy is relatively high. With all that there is little to read and no network of libraries.

There are community schools maintained by the various island communities served by volunteer teachers (mainly priests). Teaching relates to the local possibilities and requirements, but the main emphasis is on religion. Since 1980 the government has begun to wage war on illiteracy. For a rounded picture on education one has to consider the Edhuruge, the nursery school, the Makthab, the elementary school and finally the Madarasa the secondary school. At the time of writing 84 % of the population was literate. More women than men became literate as the menfolk spend much time at sea. In 1982 they hoped that by 1987 illiteracy would disappear.

If to education we added the media of mass communication, radio and TV and one newspaper, many tape recorders and record players, mainly Japanese, than we might conclude that civilisation and learning have their vessels in the Republic. True enough, but the question remains: what type of liquid passes through these vessels?

The official bicycle

Information on *reading needs* and government provisions could be gained from the following offices: education, information, planning. Their aspects are different but the general verdict is the same. Despite the high literacy rate the intellectual level was determined by a lack of reading.

Their Arabic-style writing does not easily gel into sets of sentences, let alone texts. And without texts there cannot be literature, information service, subject approach, textbooks, in brief, no such learning which depends on technical-economic progress. The knowledge of Divehi is hardly more than the use of its calligraphics. Since reading mainly serves religious purposes, writing is not taught beyond calligraphics and literature is on a minimal level – hard to come by for the lack of a library network – so progress orientated learning is not existent. This explains why the government turned to the UN Development Programme to request a plan for promoting reading in the context of a national library. I was then to prepare the ground. As a starting point the Majeed Library that had been established in 1945 was renamed the National Library in 1982. But the renaming is only symbolic, for the time being the Library is neither national nor special in any sense.

For preliminary talks I went to the National Planning Agency. I could have walked but for a tropical downpour I took a taxi which I could not dismount for the puddles, so I ended up walking into the office barefeet. This is not uncommon in Male where there are no pavements or roads with hard surface. After arrival I was offered a vehicle which I thankfully refused. It was a bicycle.

The literary horizon

Named as Majeed Library the National Library was founded by Amir Mohamed Amin Didi in 1945. The Library was assisted by the government and the users needed to have one of the following languages: Arabic, English or Urdu. The subscription costs 2 rufia (i.e. cca 20 pence) the sum was negligible the rule to pay it was not. This library still lacks both the infrastructure for learning and a functional network. Its personnel and budgetary possibilities are very limited. The money to spend on books is the equivalent of about £ 400. The staff consists of one chief librarian, two assistant librarians (one for Urdu and Maldive, one for English books), one accounting clerk, two English speaking library assistants and three general assistants. None of them have library qualifications. The accounting clerk has done a three weeks course under the direction of Mrs A. Beets in the framework of the U.S. Asia Foundation. The chief librarian Mrs H. H. H. Zubair has also attended the course and had visited the FAO Library in Bangkok.

The position with the collections is not much more favourable. The told stock is around 10,000 including the picture magazins. There are 5000 English, 1000 Arabic and 700 Urdu books. Most of the English books originate from the Island of Gan when it had a British naval base. There are sixteen serials, 13 English and 3 Urdu.

The services are curtailed indeed. There are 30 registered readers. The governmental organs have a right to use the library but they make little use of it as it lacks basic reference books. Dewey exists in an embryonic form.

There are other libraries in Male, some in a better position. The Education Development Centre Library (UNESCO-supported educational centre) is more promising in terms of stock and position. The others are mainly school and office libraries.

Lights in the night

When I arrived it was Ramadan, the period of long fasting. Walking around the streets I felt that something was strongly missing there. And then, on the last day of Ramadan when the fast had just ended, the street lit up with the lights of cigarettes. The sudden lights brought curious associations to my mind. I thought of the Arena di Verona when after the footlights of the stage are spent the spectators light candles and I thought of the Jewish habit of lighting pipes around the synagogues after the restrictions of Yom Kippur. There were other signs in Male, too. The houses began to reverberate with tape music, a change after the loud prayers of the muezzin which had been broadcasted by tannoys. Songs from Doctor Zhivago, Lily Marlene and Strangers in the Night dovetailed one another.

I felt a touch of national pride when amidst all this, I heard the tune of a Hungarian song, too, while walking among the puddles and catching the glimpse of the naked moon.

Toasting the library development plan with coconut milk

Work speeds up after the Ramadan. Meeting the Chief of Information and Radio – who was in charge of libraries, too – was pleasant and promising. It also brought the promise of a visit to Thuludhoo, which was an island of the Koafa-atoll, where a new school and an experimental industrial station were being set up.

I gave the chief my sketch plan for libraries which contains legislation regarding reading, necessities, tasks, supervision, functions, the real bases of a national library including its building and equipment. All scaled to the provisions of the country. He then asked for the plan for library legislation: urgently. At last we agreed that it will be a part of my UNESCO report.

We agreed that in order to develop the habit of reading national literature, book and periodical publication should also be developed, as well as the Divehi language to be able to express modern concepts. A system of education should connect the national plan with the library chain. We agreed that education is the crux of the matter, other things such as expert literature and equipment are easier to get hold of than professional librarians. Since they had no professional librarian in the country I recommended that there should be local courses. It is better to bring in foreign experts to the Maldives for on-job training than sending people away to learn. The Minister of Education agreed with us and we toasted the ideas with coconut milk.

So what was it we drank to? To the habit of reading, organized reading in the context of libraries and to the new professional librarians who will enforce the new laws on libraries.

The sketch of library legislation

The law should help to form a guiding body that should look after a national system for libraries. I called it *Committee for Library and Information Resources*, with the aforementioned Chief as its head. One of his officials could be the Secretary while other members should include the Director of the National Library and the representatives of the Ministry of Education, the UNESCO National Committee, the Centre for Educational Development and other bodies of learning as well as a representative for the Planning Council, the Language Council, the City of Male – all in all about 15 people. This committee should discuss the development plan of the National Library in the context of the country's needs and then advise the government on what was to be done.

The legislation's aim is to make a legislative framework for the library and information development including the role of the National Library as a place of learning. Such a piece of legislation should contain the following:

(1) Introduction

Why is the new law necessary? The specific role of the National Library in the intellectual life of the country. Three functions: national, special and public library.

(2) the National Library's system of functions

(a) Copyright for "Maldiviana" with the usual collecting range; national bibliography

(b) National Information and Reference Centre; supply of the government and other expert bodies and individuals with information that includes foreign sources

(c) a centre to coordinate the libraries of the country, applying rules and methodologies

(d) central supplies: cataloguing, exchanges of publications, including loans

(e) deposit library of international organs, U.N. and its agencies

(f) library for national lending: organizing loan facilities country wide; mobile libraries, public lending system

(3) the structure and organization of the National Library

(a) administration: policy, planning, budget, staff

(b) organizational chart: order and reference, documentation, technical services

(4) the organization of collections

(a) printed matters (books, serials)

(b) special collections (MSS, old and rare books)

(c) "gray literature" (eg. research reports)

(d) micro-publications

(e) audio-visual material

(5) reader services

(a) reading rooms

(b) borrowing

(c) reference

(d) international library loan

(e) publications, eg. national bibliography

Having passed the library legislation the first task will be to set up the rules of the National Library in accordance with the planned place of development. But it is important to note, as indeed this was the lesson I learned in the Maldives, library planning must be harmonised with local traditions of learning and in line with local economic possibilities. No model can show this except experience.

How should library legislation be brought about?

A planning group is the first thing that sets up the legislation, plans the library and participates in local induction.

The staff of the National Library might be:

1 director

1 administrative assistant

1 accounting clerk

3 professional assistant librarians (copyright and deposit functions)
including 1 children's librarian

2 professional assistant librarians (methodological and library network)

2 technical assistants (binding, conservation)
1 technical assistant (photocopying and microcopying)
2 assistants for the stores
2 general assistants

The networking librarian should organize the libraries of the islands, cooperating with their schools and cultural centres. The libraries would have a rotating stock of cca. 150 volumes in every 3 month.

Stock building needs money and organization, specially relating to exchanges. It would be possible to inspire book production by way of duplicating existing Divehi literature. The growth could be cca. 300 annually, 20 % being reference works. Serial literature around 50.

The problem of space looms large at the start. About 400 m^2 are needed right away and eventually 1500 m^2. Equipment should be there at the start.

To bring all this about the country needs international aid, bilateral and organizational, it needs foreign experts to take part in all the stages of the work.

The opening of a new school and the islands

As I mentioned above I took part at the opening of a new school and talked with the Minister of Education present about my plans. The trip was something special. As neither the capital nor the islands have proper harbours one has to change ships according to the depth of the sea. The new school was opened with suitable ceremony: ribbon cutting, treeplanting children in uniforms, parents in attendance.

This island, like Male should adopt the motto: No dogs – plenty fish. For fear of rabies there are no dogs. It is strange to be without one dog barking. Fish are plentiful like the corals and shells in all colours of the rainbow. The bather is surrounded by miriads of coloured fish which provide the staple diet. This is vital because – except for tropical fruit – there is no soil. Cooking and baking fish is spiced with the other staple diet foot: coconut. These two products are at the base of the natural economy formerly available to all. (Money if you have it could buy other things.) The third item of the diet is rice which is imported. Conserves, cool drinks, powdered milk is kept in shops mainly in the capital. The national income is based on fishing (foreign trader fish and process there), shipping and tourism. Tourist islands are now in the making, the clientele being organized by foreign hotel chains who present an earthly paradise to thousands of West European tourists.

Furasa, Vilivaru, Villingili are islands that chime like bells. On one of them I was queuing up for food and, not understanding the local lingo, tried out my own. I have kept on with this linguistic experiment until one of the islands yielded a Hungarian couple from Yugoslavia. We celebrated our meeting with fresh cold water which tasted better – after the chemical drinks diet – than French champagne.

The sea is inexhaustible in its treasures: corals must not be fished out of the sea though. However I collected a small piece whose shape reminded me of the Burghers of Calais by Rodin. Otherwise the country has some monuments and art treasures. There were many conquerors: Arabs, Ceylonese, Indians, Portugese, Duth, English, the Arabs being the most remarkable. They converted the Maldivians who were Buddhist to Islam in the 12th century. Islam's influence is ever present since, despite the fact that the intellectuals speak English. The Maldives has had a special status in the British Commonwealth.

Hello two forint-piece

There are two really very interesting sights in the capital: the main mosque and the presidential palace. The landscape is dominated by the sea and the pleasant aspects of palm-trees. The sight is of little value at monsoon times which in intensity can only be paralleled by human passion. At such occasions when it is difficult to get out, there can be, according to Mme de Staël, two ways to pass the hours: one is playing cards.

The museum is full, laden with historical and folkloristic objects. Some museologists recently finished a study on updating the museum. There is a sports stadium which is relatively modern. Here I witnessed the Independence Day celebrations in July with military parade and exercises. These were executed by the national guard to the accord of Verdi's Rigoletto.

As my time was growing to a close I had to draw a final account of my experiences. I went once more to the Centre aided by UNESCO, met with U.N. experts, talked with the Head of the local Islam Development Bank and the librarians (all three women) they shared my views. We settled that certain suggestion will be sent on location and budgeting to the authorities and what contacts the National Library must have for organizing exchange of professional information. We also agreed to write to each other.

During my last evening when packing up, a little Hungarian coin popped up among my things and rolled around the table. Hello two forint piece, after so many foreign impressions you are a token of home.

NOTE

Based on the author's account published in Magyar Nemzet 7, 14 May, 3 July 1983. Cf. UNESCO. Republic of Maldives. Development project for the National Library. Report prepared for the Government... Technical Report. PP/1981–1983/5/10.1/05. Serial No. FMR/PGI/179. (Rózsa) Paris, 1983. UNESCO 17 p.

In: TMT, 1983.8–9. 360–365.p.

THE GREAT ILLUSION OF THE MID-SEVENTIES

It is necessary to demistify the role of computerized information in development

No plan exists today for an information system, be it formulated by international, intergovernmental or nongovernmental organizations, which is not aimed at the developing countries, apart from certain regional projects and bilateral technical assistance projects. It is difficult to appreciate the results of these undertakings, and it could not be otherwise, because without a philosophy or general policy on information, everybody goes into the battle separately and in disorder.

One might mention dozens of bodies at different levels of international cooperation, which formulate policies, plan actions and administer them each one on their own, without any coherent coordination. The situation is like a badly strung tennis racket that has to be restrung to make it easier to handle, stronger and more elastic – in short, more efficient. It therefore seems that six years after having been formulated, the statement in the Jackson report is still valid: "... each of the organic services of the UN, and some specialized institutions, plans and administers its own programmes without worrying particularly about the programmes of other bodies."[1] We would point out that the information programmes for the development of specialized institutions are financed by their regular budgets and by UNDP.

Many experiments, however, have been carried out since technical assistance began to play its part in the information field. It would be very useful to summarize them because they might be the foundation of a new conception of information for development. In any event, one must first of all reconsider the general information policy of international and intergovernmental organizations, and starting from there, the policy of information for development. Regarding the latter, several projects are already operational and others are in preparation, but they concern only executive bodies and not the policy itself. It would be a serious mistake to confuse these two aspects of the problem.

Among the more ambitious and probably more promising general programmes, are UNISIST, or World Science Information System, an international programme managed and sponsored by Unesco (formerly also NATIS National Infrastructural System of Libraries, Documentation and Archives); the DEVSIS system, or World Information System for Social and Economic Development, started by the International Development Research Centre (Canada) and supported by various national and international organizations. Mention may also be made of certain sectional systems, such as AGRIS/FAO (International Information System for the Agricultural Sciences and Technology);

INIS (International Nuclear Information System) of IAEA (International Atomic Energy Agency) and UNDIS (an industrial information system) of UNIDO (United Nations Industrial Development Organization).

A real policy

However, the manner in which decisions are taken and interorganizational coordination, sometimes within the organization itself, leaves much to be desired because a general and coherent information policy is lacking. Two examples should be sufficient to illustrate this. The Inter-Organization Board of Information Systems and Related Activities (IOB) has been set up to coordinate the computerized information programmes of specialized agencies. This same body has been placed under the guardianship of the institutions that it was supposed to coordinate and by which it is for the most part financed. This therefore implies a high-level self-administration since the employees must be capable of coordinating the activities of their employers. The two information programmes had been started by the same organization within three years of each other, in this particular case the above-mentioned UNISIST and NATIS in the two programmes in question had been incompatible. They were administered by divisions belonging to different departments of the organizations (observation in 1987: the PGI is the unique and global information programme of UNESCO).

So what would be a real information policy? First of all, its administration should be global and coherent, in order to transform all the organizations involved in information activities into an actual network, the members of which would follow a common policy. To change the situation one would need the creation of a body that would be a kind of concultative committee, working together with government experts, and an information bureau for development, which would be made up of information, planning and administration experts.

Centralization of decision

The information bureau, which would handle economic and social information and whose field of activities would extend to planning, operational projects and structures, would exert a certain control over the ordinary budget of the United Nations and its system concerning the library and documentation services and programmes, as well as over the extra-budgetary funds (assigned to information technical assistance).

Such method would allow the exercise of a real influence over the carrying out of information programmes and projects, especially of the information programmes for development. It would then be possible to combine effectively a centralization of policy, with the decentralization and enactment of projects by specialized institutions or by other international or national organizations. J. Tocatlian (Unesco) in his *International Information Systems*[2] analyses dozens of programmes, intended to be regional or world-wide, launched by national or international organizations, based mainly on information systems established and developed independently of each other. Meanwhile, between 1946 and 1973, UN documents showed no less than four hundred references to information policy, and coordinating information for development. So efforts have not been lacking. (Observation in 1987: it can be considered that between 1973 and 1987 another hundred references became available concerning information policy for development and for the coordination of these policies. In this respect Unesco took a very important decision by establishing its global information programme, the PGI-Programme Général d'Information through the combination of UNISIST and NATIS. Among the 5 major orientations of the PGI's one concerns the information for development. This step can be considered as an attempt for the integration of information policy for developing countries. However this attempt for integration concerns mainly Unesco's activities in the matter of development information. It shows the suitable tendency of international information policies for development.)

However it seems certain that without concentrating the powers of decision on matters of policy and financial administration[3] – a very real power – it is difficult to achieve appreciable progress in the development of information and, above all, in information for development.

This centralization would certainly be a great step forward; it should, however, be accompained by a serious consideration of the actual role of information in the development process. In this connection, there are many misunderstandings, overestimates and myths. There are many reasons for this phenomenon. First of all, it is felt that the political authorities should be sensitized to the problems of information, especially in the national field, so that eventually they may provide the necessary means to solve them. And who is in a better position to provide the subject for this continuous campaign than the international bodies specialized in information? To render services to developing countries is one of the reasons for their existence. The work of these bodies is reinforced by that of the big computer manufacturers, who consider themselves to be indispensable in all fields of information.

No doubt that scientific information brings an influence to bear on social structures, but it is different factors, the economic and social factors, which influence the structure of society.

It can be concluded that, even while continuing the process of sensitizing the national and international authorities with the aim of supporting in all possible ways the planning of scientific information services, the role of computerized information should be demystified, especially in the interest of developing countries. One must not lead people to believe that to organize services of this kind in those countries, even with the most sophisticated systems, will have a primary influence on their development.

By sophisticated services, we mean computerized information systems. The idea of "information revolution," with the computer as the panacea, has contributed to confuse the issue and to dissimulate the real problems of information for development.

Sophisticated techniques

This confusion has already been sufficiently costly for the countries concerned and will cost them even more. However, it is not the computer in itself that is expensive, but rather the illusion that with it, everything will be better. Instead, it can lead to a diversion. This diversion leads to great waste, as it always happens when the means employed are clearly superior to the conditions in which they are or will be used. This means that the use of these sophisticated techniques is justified only if the situation is really favourable, which is rarely the case. It is not the computer that brings about development, although at a certain stage of expansion, an economic and technological infrastructure may require a changeover to a higher technical level. In these conditions, information can then become an extremely powerful instrument for future development. Conversely, if the time of this changeover is not well chosen, a situation occurs in which the developing countries will more than ever depend on the countries providing these advanced techniques.

Not a cure-all

Yet, it is not the form but the content that makes information "modern." Furthermore, if there is one thing that the developing countries have to keep in mind with regard to computerized information, it is its impersonality and the fact that it can't replace the use of primary literature, that of reading. The reading of specialized literature is a means of training and recycling. For those who are able to make an effective use of it, the computer is a major tool, but it could never be a cure for all evils. And, generally speaking, this goes for scientific information services, which can make only a limited contribution to speeding up development.

To decolonize information, it is advisable therefore to clarify the role it plays in development. This is the first step to take. Then one must admit that there would be more than one road to follow and that there is no general pattern for all countries. Since every case is special, the policies, methods and technologies to be followed must be "made to measure" for every country or every region. Every country should be responsible for its information policy taking into account international experience and the experience of countries with similar conditions.

Information for development, if well understood, should give absolute priority to social and human factors, information media, including technology, being subordinated to these factors. Training should have first priority.

It must also be remembered that the application of advanced technology (informatics techniques) in unsuitable conditions (economic, technical and sociocultural underdeveloped infrastructures), does not lead toward independence, but on the contrary, toward dependence.It is therefore, according to the stage reached in institutional maturity and in complexity of public administration and company management, according to the training level of personnel, scientific and technical works, that the methods and technologies of information should be chosen. Social and human information plays an important part in development, at three different levels: *formation of the national conscience* (history, ethnology, linguistics), *long-term economic development* (teaching, professional training, general culture), *short-term development* (applied economics, management).

Let us also remember that in developed countries the volume of documentation is the consequence of the abundance of knowledge and of the uncontrolled flow of publications in industrialized countries. This does not happen in developing countries, where the information services must operate in conditions of scarcity. It is therefore essential to ensure at least minimum access to documents, and to create *library services* that will meet real needs.

From what has been said above, it follows that one must look for new ways and methods of information for development. Scarcity (like excess) of information creates the conditions for dependence. The division of labour on which the information systems of the developed countries are founded may not be as indispensable as in the developing countries, owing to their present conditions. Should advanced specialization be the road to follow? Should it not be possible to consider information for development as an integral part of the activities at the service of social progress instead of breaking it up into specialized programmes (agriculture, industry, etc.), which are hardly compatible?

Could not one imagine for example having information units of a *new kind of know-how*, something like *clearing houses of applied information* (information-extension-training) comprising the following units:

– a reference service with collective catalogues, basic works, encyclopaedias, magazines,

– an automated information service with its terminal connected to the big teledocumentation systems, equipped with teletypes and copying machines,

– an audio-visual demonstration service with films, recordings, exhibits and other extension methods, etc.,

– a "know-how" service including workshops, farm schools and experimental stations for adult training and refresher courses,

– a pedagogic and consultation service to prepare the pedagogic know-how material (production and marketing guides, etc.) and distribute it, to prepare the course programmes combined with demonstrations and production and give counsel and advice,

– general services with the function of financial administration of the centre, including a laboratory for applied psychology and professional orientation with integrated economic and sociocultural analysis.

For the common interest

In other words, all these services together, which could be adapted to meet different needs, would constitute a multi-functional and multidisciplinary centre of applied information, equipped with knowledge and know-how, where *research would be combined with information, training and work.*

If one could get rid of preconceived ideas, prejudices, the craving for prestige, feelings of superiority and inferiority, if one could break loose from sectional interests for the benefit of common interest so as to give free rein to creative imagination, one could then find new formulas, expand the frontiers of information for development, transforming it into applied information, which is tied to social progress: the real reason for the existence of information services.

NOTES

1. Étude de la capacité du système des Nations Unies pour le développement. UN, Geneva, 1969.
2. Advances in Librarianship. Vol. 5, 1975, p. 2–52.
3. On this subject, see A new United Nations structure for global economic cooperation. UN, E/AC.6219, 1975.

In: Ceres, FAO Review on Development, 1975.5. 35–38.p.

MODELS FOR THE INFORMATION INFRASTRUCTURE OF DEVELOPING COUNTRIES

1. What about information infrastructure?

The concept of infrastructure covers in the broad sense those social-economic, cultural factors and services which – although they do not directly produce goods themselves – influence material production. Social science literature of recent decades – and especially the literature of economics – has dealt in great length with the primary role of the infrastructure in social-economic progress. At least two motives of this increased interest should be recognized. One is that as a result of technical progress, less and less people participate in material production while the number of people engaged in service activities grows rapidly. These services also affect the cultural sphere, the organization of utilizing leisure time and the most diverse programmes linked to this area. The other motive is the complexity of problems related to the progress of the developing countries where – due to their particular situation – infrastructure plays a crucial role.

In both of these motives the weight of information – to be more exact, the weight of scientific information – is equally significant. The literature of industrialized countries covers an emerging information society, where the formative role of information, especially scientific information, appears to be decisive. As regards the "Third World" the progress of the information system in the broad sense is a basic element in the development strategies for the spreading of literacy and education. I consider the development of this system to be a part of cultural planning (i.e. programming and budgeting of cultural activities) and shall now discuss this topic.

The fundamental question in the development of the information system of the "Third World", or to be more exact of their information infrastructure, is highlighted by the following Chinese aphorism: "If you give a fish to a hungry man, it will appease his hunger only once; but if you teach him how to catch fish, he will have enough food for the rest of his life." The up-to-date, humane answer to this dilemma, which at the same time promises an efficient solution, can hardly be in doubt.

Fish should be provided too, but the poor man should be taught to fish as soon as possible. This is how far the parable goes, but what should be done in practice? What should be done that – to transfer the allegory from China to India – an elephant head should be put on the human body? This may be called the *Ganesa-effect*. According to legend, Ganesa, the elephant-headed Hindu God, the son of Shiva and Parvati and the legendary scribe who wrote down the Mahabharata was accidentally beheaded by his father. To ease the grief of his wife, he promised to cut off the head of the first living creature and join it to the mutilated body. The first creature was an elephant.

2. Specialization or integration?

Staying with the allegories of the parables, there are certainly several ways to the teaching of fishing, several kinds of methods, and there are several kinds of elephants too. The development of the infrastructure of the "Third World" does not have a single model, but *several* models exist. The differences are not only represented by the models to be applied, but also by those who apply them. Developing countries differ from each other in many respects in their economic development, historic and cultural endowment, tradition, social structure, etc. What I consider to be *common* in all models is that the processes, techniques, training to be applied, must always be harmonized with the local capabilities and circumstances. Disregarding this requirement would result in the organism rejecting the alien body.

The selection of the processes to be applied (training, etc.) must be made with consideration of the entire local situation. This refers to the technical assistance programmes in general, and among them especially to the programmes directed at the development of the information infrastructures.

After considering the local situation the following working hypothesis may be applied in the course of planning these programmes, whether the programmes be of a bilateral or multilateral nature or those to be implemented by international organizations. To what extent should the model of infrastructure be of an *integrated* or a *specialized* nature? In my opinion, the less favourable the socioeconomic and cultural conditions are, the less useful it is to rely on specialization. The obvious reason for this consideration is that limitations in the available expertise and technology make the consideration of human and material resources necessary. The creation of an information infrastructure can be hardly succesful if the sectoral activities (industry, agriculture, transportation, etc.) on the one hand, and the division of institutional activities (such as education, libraries, archives, documentation, etc.) on the other, were overspecialized. In some cases it appears justified to develop an information infrastructure of an integrated nature which includes the elements of several or of all the sectors above, as well as the various cultural institutions, such as the special library, public library, documentation centre, archives, whose responsibilities include the organization of reading courses and publishing as well. This applies especially to countries where there is, on the average, no more than one specialist for each information institution. And what can be done for instance in countries where only ten specialists are available for all the information institutions? In such cases the various sectoral and institutional information activities should be highly integrated. The degree of integration has to be interpreted flexibly: in time, according to need, some functions can be expanded, or made independent. In the least developed countries, which are at the bottom of the economic ranking order according to the per capita income, it is, in all probability, advisable to apply the integrated model.

3. The so-called BACIN-Model

Such an integrated model, which we call BACIN (from the French initials of *B*ibliothèque Nationale, *A*rchives nationales, *C*entre national d'*In*formation) should consist of the following components:

	General Directorate Secretariat, general management	
Library (national, special, public)	Archives	Scientific information
Acquisition and processing (catalogues)	Pre-recording	National bibliography and special bibliographies (BACIN-publications)
Reader's services (reading rooms, references)	Archival Group	Computerized data bases
Central registers (ISBN, ISSN, union catalogues)		Information syntheses (state-of-the-art reports)
Research, methodology (e.g. networking, standardization etc.) and training		Technical development (data processing and other mechanization)
	The united technological base (laboratories, workshops, etc.) covering the three branches	

Table 1.

I believe that the essence of the information infrastructure of the integrated type can be understood from the BACIN-pattern. The three basic branches of information activities, – library, archives and documentation perform diverse tasks within the same organizational entity; their logistic arrangement and division of labour underlie their special character. This scheme is probably the most economical in terms of manpower management and operational expenditures. Naturally, additional functions can be linked to the activities represented in the pattern. It is also possible to simplify the scheme by omitting certain operational elements or merging them as necessary in special circumstances. For instance, there are countries where only a very limited national literature exists and the purchasing power is inadequate. In such countries it would not be easy to sell books. In such circumstances, and for library use only the already existing works of national literature should be multiplied by a BACIN-type institution through a cheap process. The same applies for some pieces of foreign language fiction or non-fiction literature. The circulation of these works should be organized by the library network. In fact, this library network should possess a central stock that could be regularly rotated by the main institution. This way an institution of the BACIN-type may see to it that the already literate sector of the population should not neglect reading for the lack of reading matter. The institution may also be engaged in post-training in reading.

4. The most important: the human factor

The training function however, should not be limited to reading programs; it may assume an active role in public education and professional training. I am thinking here, for instance, of the training of librarians and information specialists, because in many countries the conditions do not exist for establishing such specialized training institutions. It appears useful to organize the training of librarians and information specialists *in a national framework* with the assistance of foreign guest lecturers. It is only worth sending people abroad who possess sound basic knowledge upon which specialized training can be built. Another advantage of training within the country is to connect theory and practice in a realistic way. Practices which can be learned in advanced countries cannot always be applied in the national framework of developing countries due to their lack of facilities and instruments. The consequence of this may be, a low level of efficiency, the frustration of the "overeducated" information specialist, and a number of other negative results. The desire to apply advanced foreign practice which are doomed to failure in an inadequate and resisting home situation very often results in deep conflicts.

5. High technology and reading

It is another source of conflict if the most advanced information technology was applied without the foundations of reading, the availability of reading matter without the specialized know-how. For the successful adoption of technology transfer an adequate level of cultural preparedness is needed as well as the availability of a collection of primary documents. It is on these grounds that computerized data bases can be built. It is necessary to learn about mechanization, to experiment with it, but never with the wrong assumption that information technology may be a substitute for primary literature. Only step by step progress, the patient building of foundations and support-structures will strengthen technology transfer. In the integrated model of the BACIN-type, or in other similar models, the availability of documents and the computerized data bases *complement each other*. It takes hard work to climb up the tree of high technology, it is not possible to jump on it like on a slow-moving carriage. The specialized information infrastructure is at present the universal one. The reason for this is, first, that all examples seen by the developing countries are based on specialization. The transfering institution – be it national or international – is itself strongly specialized.

Secondly, the institutional system of developing countries also follows the course of specialization and this trend attracts the sectoral and functional areas (industry, agriculture, national library, special library, information centre, education, etc.). But the many requirements posed by this trend – e.g., the establishment and housing of collections, manpower and equipment, etc. – are very difficult if not impossible to satisfy. In industrialized countries these demands are expected and met, but not always adequately. Yet in the case of developing countries specialization results in the fragmentation of manpower, dwindling funds because of a weak financial potential and specialists limited in number. Despite these considerations, the development of specialized models will continue since – as I have already mentioned – this kind of model is dominant in advanced conditions. Apparently, the United Nations' family progresses in the same direction.

Why is it then still worthwhile to outline the features of the integration model when, according to all indications, such models have little chance to be adopted by administrations orientated towards specialization? The reason is the anticipation that infrastructural developments based on integrated models will be found attractive to the least developed countries (LDCS). Besides the numerous attempts of developing information infrastructures based on specialization, the trial of an integrated model may also occur. There are already recommendations of this nature in specialized literature, and international organizations and governments have also received expert reports advocating this route (including some by this author).[1]

It would perhaps be worthwhile to try them *as an experiment* in two or three countries to see what the results of the application of integrated models would be. It is to this aim that I wanted to contribute some ideas.

NOTE

The present author has made use of the two UNESCO Technical Reports prepared by him. UNESCO. Republic of Maldives. Development project for the National Library. Report prepared for the Government... Technical Report. PP/1981–1983/5/10.1/05. Serial No. FMR/PGI/82/179. (Rózsa) Paris, 1983. UNESCO 17 p. UNESCO. Rwanda. Le développement d'un système intégré d'information scientifique (BACIN). Rapport de mission PP/1981–1983. No. de série FMR/PGI/83/189. (Rózsa) Paris, 1983. UNESCO. 21 p.

In: Information, Communications and Technology Transfer. Ed. by E.V. Smith and S. Keenan. Elsevier Publ. FID. 1987. 15–19.p.

TECHNICAL NOTES AND ABBREVIATIONS

The three topics of this collection of papers form a thematic unity: information claims, the economics of scientific and scholarly information and international cooperation in information are presented together. But because these studies, essays and articles span two decades and large distances in location the individual publications reflect a movable picture. Consequently, no effort was made to bibliographical consistency in description, transliteration or capitalisation. These are formal variants which appear in the volume as they had appeared in the original publications.

When it came to editing the author concentrated on updating some of the notes, references and augmenting certain passages of the text when this was deemed to help the user.

The majority of these studies appeared in English in international organs devoted to the subject. The rest, which had appeared in French, Russian and Hungarian were translated into English. The original sources are given at the conclusion of the individual articles.

A list of selected abbreviations may facilitate the use of the volume:

ACAST	Advisory Committee on the Application of Science and Technology to Development
ACC	Administrative Committee on Co-ordination
ACCIS	Advisory Committee for the Co-ordination of Information Systems
AGRIS	International Information System for the Agricultural Sciences and Technology
AIC	Applied Information Centre
ASCA	Automatic Subject Citation Alert
ASIS	American Society of Information Science
BLCMP	Birmingham Libraries Cooperative Mechanisation Project
BSO	Broad System of Ordering
CARIS	Current Agricultural Research Information System
CDS	Computerized Documentation Service
CESI	Centre of Economic and Social Information
CIPMD	Committee on Information Planning and Management for Development
CLADES	Centro Latino-Americano de Documentación Económica y Social
CMEA	Council for Mutual Economic Aid
CNRS	Centre National de la Recherche Scientifique
COMECON	Council for Mutual Economic Aid
COPRO	Common Processing on Computer Processed Data

CORE	Common Register on Development Activities
CSTD	Committee on Science and Technology for Development
DARE	Data Retrieval for Social Sciences
DEVSIS	Development Sciences Information System
DHL	Dag Hammarskjöld Library
ECE	Economic Commission for Europe
ECOSOC	Economic and Social Council
ECSSID	European Co-operation in Social Science Information and Documentation
ECSSID/IOC	ECSSID International Organizing Committee
ECSSID/WG	ECSSID Working Group
EDP	Electronic Data Processing
ERIC	Educational Resources Information Center
ESRO	European Space Research Organization
ESRO/RECON	ESRO Remote Console
EUDISED	European Documentation and Information System for Education
EUSIDIC	European Association of Information Services
EUSIREF	European Referral Service
FAO	Food and Agriculture Organization
FID	International Federation for Documentation
FID/RI	FID Committee on Research on the Theoretical Basis of Information
FID/SID	FID Social Development
FRANCIS	French Retrieval Automated Network for Current Information in Social and Human Sciences
HAS	Hungarian Academy of Sciences
IAEA	International Atomic Energy Agency
IASSIST	International Association for Social Science Information Services and Technology
ICSSID	International Committee for Social Sciences and Information Documentation
ICSU	International Council of Scientific Unions
IDRIC	Industrial Development Research Council
IFDO	International Federation of Data Organizations
IFLA	International Federation of Libraries Association
IGOs	International Intergovernmental Organizations
ILO	International Labour Organization
IMF	International Monetary Fund
INDIS	Industrial Information System
INFOTERM	International Information Centre for Terminology
INION	Institute of Scientific Information on Social Sciences
INIS	International Nuclear Information System
INPADOC	International Patent Documentation Center

INTERCONCEPT	International Information System of Social Science Concepts
IOB	Inter-Organization Board for Information Systems
IRS	International Referral Service for Environment
ISDS	International Serials Data System
ISIS	Integrated Science Information System
ISORID	International Information System on Research in Documentation
ISSC	International Social Science Council
LDCS	Long distance control system, datapoint
MISON	International Social Science Information System
NATIS	National Information Systems
NGOs	Non-governmental organizations
OECD	Organization for Economic Co-operation and Development
OST	Office for Science and Technology
OTID	Office for the Transfer of Information for Development
PGI	General Information Programme (Unesco)
R + D	Research and Development
SCANDIA	Scandinavian Common Accession Programme
SDI	Selective Dissemination of Information
SOINES	Science and Technology Policies Information Exchange System
SSCC	Social Science Coordination Committee
SSID	Social Science Information and Documentation
SWALCAP	South West Academic Libraries Cooperative Automation Project
SWIDOC	Social-Wetenschappelijk Informatie en Documentatie Centrum
TMT	Tudományos és Műszaki Tájékoztatás (Scient. and Techn. Inf.)
UDC	Universal Decimal Classification
UN	United Nations
UNBIS	United Nations Bibliographic Information System
UNCSTD	UN Conference on Science and Technology for Development
UNCTAD	United Nations Conference on Trade and Development
UNDEX	United Nations Document Index
UNDP	United Nations Development Programme
UNEP	United Nations Environment Programme
UNEP/IRS	UNEP International Referral Service
UNESCO	United Nations Educational, Scientific and Cutlural Organization
UNHCR	Office of the United Nations High Commissioner for Refugees
UNICEF	United Nations Children's Fund
UNIDO	United Nations Industrial Development Organization
UNISIST	Universal System for Information in Science and Technology
UNITAR	United Nations Institute for Training and Research
UNOG	UN Office at Geneva Library
VINITI	All-Union Institute of Scientific and Technical Information
WG	Working Group
WIPO	World Intellectual Property Organization

INFORMATION

Contents

AUTHOR'S BIOGRAPHY

György RÓZSA is a chartered librarian, Doctor of Economics, 1960– Director General of the Library of the Hungarian Academy of Sciences, meanwhile, from 1969–1975 Director of the United Nations Library in Geneva. He worked 1946–1950 in the foreign service, from 1951 in the library field, and since 1954 with the Hungarian Academy of Sciences.

Works: Sources and Handbooks of Economic Research (1959); Information Problems of Social Science Research and of Research Organization (1965); Scientific Information and Society (1972); the latter was published in English as well by Mouton in 1973 (Paris – The Hague). Gy. Rózsa published nearly 350 articles, studies and papers in Hungarian and other languages on the subject of scientific information and research organization (ref., LISA, ERIC, Bull. Sign., Könyvtári és Dokumentációs Szakirodalom, etc.). He also worked as consultant commissioned by international organizations, and is chief editor of two specialized information journals: *ECSSID Bulletin*, an international newsletter and *Research and Development*, a research management information journal (in Hungarian). He participates in the work of Hungarian and international professional bodies (e.g. FID/RI Committee, ECSSID Executive Committee).

Professional awards: Special Medal of the International Council of Archives (1977), "Szabó Ervin" Memorial Medal (1981).

Felelős kiadó: az MTA Könyvtára főigazgatója
Alak: B/5 – Terjedelem: 18,4 (A/5) ív
Megjelenés: 1987 – Példányszám: 800
Készült az MTA Könyvtára házi sokszorosító
részlegében

libro absolutamente carente de interés, salvo por cuanto refleja hasta la caricatura cierta producción de paja en Takeoff (aparente o real): pedantería, ciencia extranjera mal asimilada, confusión inelegante; en fin...